QUANTUM MECHANICS

$$\Psi_{Life} = \frac{1}{\sqrt{2}} (\uparrow_{success}\downarrow_{failure} - \downarrow_{failure}\uparrow_{success})$$

Basic & Advanced Concepts for Beginners

Preetinder Rahil

Limit of Liability/Disclaimer of Warranty

The contents of this book are generic in nature and not directed at anyone specific. The book is not intended for academic purpose. The readers are advised to consult standard textbooks for that purpose. While the author and publisher have made best efforts in preparing this book, no warranties to the accuracy and completeness of the information in this book are made. The internet links provided may not be accurate or appropriate over time, the readers are advised to take due diligence. Neither the publisher nor the author shall be liable for any loss, liability or risk related to the contents of the book, directly or indirectly.

Copyright©2018 by Preetinder Rahil. All rights reserved. No part of this publication may be reproduced, distributed or transmitted in any form or any means, or stored in a database or retrieval system without the prior written permission of the publisher except as allowed under the Copyright Law.

Table of Contents

Preface

Conversation Starter

Chapter 1 - Review of Mathematics	11
Chapter 2 - Classical Mechanics	22
Chapter 3 - Laws of Quantum Mechanics	36
Chapter 4 - Simple Quantum Models	89
Chapter 5 - Hydrogen Atom	118
Chapter 6 - Spin	147
Chapter 7 - Quantum Computing	186
Chapter 8 - Identical Particles	202
Chapter 9 - Approximation Methods	239
Chapter 10 - The Standard Model	338
Chapter 11 - Einstein's Special Theory of Relativity	347
Chapter 12 - Relativistic Quantum Mechanics	370
Chapter 13 - Strong & Weak Interactions	412
Chapter 14 - Quantum Field Theory	438
Chapter 15 - Beyond Quantum Mechanics	453

References 487

Index 490

Preface

Quantum mechanics is one of the most fascinating subjects. How the nature behaves at the quantum level pales even science fiction. If the strangeness and weird behavior at the quantum level manifests in our daily life, it would become unrecognizable. It is difficult to say what is more surprising, the weird quantum mechanics or the human mind that can come up with the theory to explain it. Quantum mechanics is not just another subject. It tells us how universe began and what will be the fate of the universe. There are deep philosophical and religious questions that emerge from the study of quantum mechanics. But we cannot have any serious discussion about the implications of quantum mechanics without understanding it first. It is unfortunate that more than 100 years since the quantum mechanics revolution started, very few people outside of physicists have any working knowledge of the subject. It is a tragedy as we are taught subjects like classical mechanics in high school and basics of quantum mechanics ought to be taught at the school or college level. There is no dearth of appetite among the public to learn quantum mechanics. Quantum phenomenon like the uncertainty principle and Schrodinger's cat have become part of the popular culture. There are several popular books that have been written and documentaries made to explain the quantum weirdness. So, what's the point of writing another book?

This book is not a popular book in that sense. Its content is closer to a textbook of quantum mechanics. However, it is not a typical textbook either. The popular books do a good job in explaining the basic phenomenon, but they do not go in detail of the equations or the mathematics behind the subject. It's like telling you how beautiful and exceptional a car is to drive without taking you for a test drive. You do not get the real flavor of quantum mechanics. On the other hand, quantum mechanics textbooks are too hard to understand for a non-physicist. The mathematics is advanced and rigorous, and it is assumed that you have a good understanding of advanced mathematical techniques like solving differential equations and complex functions. There is a good deal of emphasis on problem solving and exam-based material, which is needed if you want to get a university degree. But it is of less interest to general public who just want to get the flavor of quantum mechanics. In my quest to learn quantum mechanics, I found there is a gap between popular books and textbooks. There is need for a middle ground, where details of equations and mathematics involved are explained without going into the didactic requirements to pass the exams. You must be thinking that this idea must have come to other people as well but why there is little action on it. I think it is just too hard to write such a book and there is no compelling commercial reason.

What credentials do I have to write such a book? I am not a physicist but a physics enthusiast. Whatever I have learned is through self-study. I am a physician by profession and usually physicians tend to run far away from anything mathematics related. It should be huge disadvantage to write on such a difficult subject. But I believe, it is an advantage. I am quite familiar with the problems faced by a non-physicist when trying to learn the subject. I assume no prior knowledge of any mathematics except a bare minimum, which a high school graduate ought to know. I have tried to explain the mathematics on the go. The mathematics is easier to learn in a context and I have tried my best not to take short cut when solving equations. I have used unconventional techniques to give intuition behind core concepts and mathematics. I believe if you can visualize the phenomenon or at least relate it to our daily life, the learning is much faster. It is easier said than done and many physicists believe that in case of quantum mechanics, it is a tall order. All

I can say is you can judge it for yourself if I am successful in explaining you the key concepts. I have liberally used analogies from the world of finance to politics which a professional physicist will be little anxious to use. But these are analogies to explain the phenomenon, nothing more. Do not take them literally!

The material included in this book is at the university level. The topics are in fact a lot more than what is included at a beginner's course in quantum mechanics. I have included additional topics from relativistic quantum mechanics to quantum computing, from quantum field theory to string theory. Try not to memorize equations or get stuck at learning every step of the equations. Get the feel of the subject and develop intuition. It may require reading the book many times. Once you get the feel of the subject, try reading a textbook like Griffith's or Shankar if you are feeling brave.

I am thankful to all the sources that helped me learn the subject. I have included the list at the end of the book. I welcome any suggestion to improve the book. Please do not hesitate to contact me if you find any errors or inaccuracies in the subject material. I hope you will enjoy reading and learning quantum mechanics from the book. A word of caution for students, this book is not designed to replace textbooks or be used for academic purposes. I have tried my best to check the accuracy of the equations but please check with your standard texts before you use any material for academics.

Happy reading!

Conversation Starter

Quantum mechanics is a serious subject. It requires focus and dedication. But we will start on a lighter note. Let's have a fictional conversation between a classical physicist and a quantum physicist. There is no better person than Newton to represent classical physics. The quantum physics will be represented by Paul Dirac. It is unfortunate that very few people know his name. He was one of the greatest physicists ever lived and made a huge contribution to quantum mechanics. He was an unassuming and low-key person. His persona was legendary. His contemporaries like Einstein, Pauli, Heisenberg etc. had interesting quotes about him. There have books written to describe his unique personality. We are fortunate that there are videos of him giving lecture on quantum mechanics on YouTube. He was quite old at the time but still it is a historical treasure.

Newton visits his old university department at Cambridge. He finds Paul Dirac as the Lucasian Professor of Mathematics. Newton himself occupied this post once. Interestingly, Stephen Hawking held the same post as well.

Newton: I am Sir Isaac Newton, past Member of Parliament, President of Royal Society, Master of the Royal Mint, Lucasian Professor of Mathematics and the author of Principia Mathematica. Who are you?

Dirac: I am Paul Dirac and I have an equation called the Dirac equation.

Newton: Good for you. But I have laws of motion, among other things. There is a rumor going on in the heavens that Newtonian physics is in danger and new kind of physics is replacing it. Is Newtonian mechanics wrong?

Dirac: Not wrong, just inaccurate. If you take the average results of quantum mechanics, we get Newtonian mechanics.

Newton: Classical physics is not average but exceptional. I will hear you out, tell me what's it about?

Dirac: Let's start with the basics. Suppose you have a friend, Bob. If you want to find him, he is either at home, university or in the pub. How will you find him?

Newton: Is that even a question? You look for him and find him!

Dirac: Get ready for quantum weirdness. If Bob is a quantum particle, he will be at all three places at the same time. He has no fixed position. We say he is in superposition. Once you decide to look for him, the superposition goes away, and he appears randomly at either of the three places.

$$Bob = \frac{1}{\sqrt{3}}(Home) + \frac{1}{\sqrt{3}}(Univ.) + \frac{1}{\sqrt{3}}(Pub)$$

We can only say about the probability of finding Bob, which is $1/3^{rd}$ in each case but the result is still random.

Newton: This is nonsense! How do you know that Bob was in superposition? He was always there when you found him.

Dirac: This is a good question. We have to do millions of experiments to get the data to know that the result was random and the probability of finding the Bob at each place. If Bob had a clone, and you did the same experiment, he will be found at a random place. This applies to quantum particles only like same state electrons.

Newton: What's up with looking at the Bob causes the superposition to collapse? Are you saying that you exist only if I look at you?

Dirac: It appears so. A quantum particle has no fixed position until it is measured.

Newton: If the world is that crazy and weird, why don't we see it in our daily life?

Dirac: The quantum phenomenon only manifests itself at the quantum level. Once you collect billions and trillions of particles to form classical objects like human beings, the quantum effects average out to give classical physics.

Newton: Thank God for that!

Dirac: Things get worse. Even if you know the position of the particle, you do not know where it is going. If you know where it is going, the position measured earlier is no longer valid. There is always uncertainty about the quantum particle whereabouts.

Newton: If quantum mechanics is so weird, how the hell can you do any calculations?

Dirac: We use calculus and the Leibnitz notation.

Awkward silence

(Trivia-There was a bitter dispute in 1600's about who invented calculus first-Newton or Leibnitz. The modern notation is based on the work of Leibnitz)

Newton: You don't know where the particle is, where it is going. What do you know?

Dirac: I probably know where the particle is and where it is going but I certainly know that quantum mechanics is right.

By the way, meet Einstein, he has something to say about Newtonian gravity.

Chapter 1

Review of Mathematics

I know it's not a good omen to start with mathematics. But there is no escaping it. The language of physics is mathematics and we need the necessary tools to explore the quantum world. But we will only study bare minimum mathematical facts and that too in a way which is intuitive.

Calculus

You must be thinking, I gave up science in college to avoid calculus and how am I going to master it now? The bad news is that study of calculus is absolutely essential, but the good news is that we can get away with limited number of concepts. I will only include elementary facts that's nothing more than high school stuff. The heavy-duty mathematics will be explained as we study quantum mechanics in detail. The best way to learn mathematics is to solve a problem at hand and learn the intuition behind it. Cramming mathematical formulae is a low yield exercise. Nevertheless, some basic facts will need to be memorized that will increase our speed of learning.

Differentiation

It is the study of how things change at an instant.

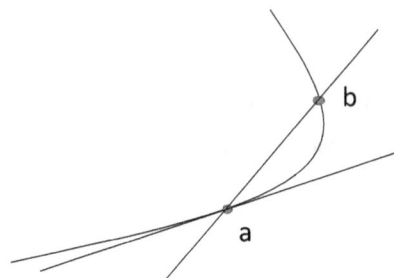

In the x-y plane, how a change in x affects y is given by the slope $\frac{\Delta y}{\Delta x}$. The change from point a to b can be calculated by calculating the slope. If we move b closer and closer to a so that it's almost at a then the instantaneous change is given by the tangent at point a.

In the calculus notation, it is called $\frac{dy}{dx}$.

Each curve will have a unique tangent or differential. The first principles are used to calculate them using algebra and geometry, but Newton and Leibniz have already given us the answers, we just need to memorize them.

Here are the basic differentiation formulae.

$$\frac{dx^n}{dx} = nx^{n-1}$$

e.g. $\frac{dx^2}{dx} = 2x, \frac{dx}{dx} = 1$

$$\frac{d\ constant}{dx} = 0$$

e.g. $\frac{d\ 5}{dx} = 0$

The logic here is simple. If something is constant, how can it change?

The product rule

$$\frac{d(xy)}{dx} = y\frac{dx}{dx} + x\frac{dy}{dx}$$

e.g. $\frac{d(x^2 y)}{dx} = y\frac{dx^2}{dx} + x^2\frac{dy}{dx} = 2xy + 0 = 2xy$

Note the derivative of $\frac{dy}{dx} = 0$ if y is not a function of x.

The chain rule

$\frac{d(x^2+1)^2}{dx}$, let $u = x^2 + 1$

First differentiate u then the stuff in it.

$$\frac{d(u)^2}{dx} = 2u\frac{du}{dx} = 2(x^2+1)\frac{d(x^2+1)}{dx} = 2(x^2+1)2x$$

Differentiation of special functions

$$\frac{de^u}{dx} = e^u \frac{du}{dx}$$

$$\frac{de^{x^2}}{dx} = e^{x^2} \frac{dx^2}{dx} = 2xe^{x^2}$$

e is a constant with value of 2.71 and it is the basis of natural algorithm. We will encounter it a lot in quantum mechanics, so remember how to differentiate it.

It represents exponential growth or decay and its properties are extremely useful. With exponential growth, as things grows bigger, growth rate increases.

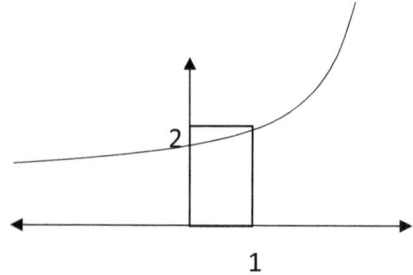

$e^1 = e = 2 \cdot 71$, $e^5 = 148$, $e^{10} = 22026$ and so on, it grows fast!

$$\frac{d}{dx} sinx = cosx$$

$$\frac{d}{dx} cosx = -sinx$$

Partial Differentiation

There can be variables besides x that we want to calculate their rate of change. The process is pretty simple, just differentiate with respect to each variable. The variable only acts on the function that depends on it.

$$\frac{d}{dx} \rightarrow \frac{\partial}{\partial x}$$

$$\frac{\partial}{\partial x}(x^2 + 2y^2) = 2x + 0 = 2x$$

$$\frac{\partial}{\partial y}(x^2 + 2y^2) = 0 + 4y = 4y$$

That's enough differentiation for now.

Integration

It is differentiation in reverse. It is going backwards from differentiation to get the original equation.

Let $y = x^2$ then $\frac{dy}{dx}$ is

$$\frac{dx^2}{dx} = 2x$$

And the integral is $\int dy = \int 2x dx$

The basic polynomial formula of integration is $\int x^n = \frac{x^{n+1}}{n+1}$

So, $y = 2\frac{x^{1+1}}{1+1} = x^2$

There is a slight problem, the differentiation of $x^2 + 1, x^2 + 2$ etc. is also 2x. So, when we integrate, which equation are we getting?

To avoid this problem, we add constant of integration whose value depends on your choice or what we call initial or boundary conditions.

$y = x^2 + K$

The value of K can be 1,2 or whatever the situation demands. This is why integration is a bit harder than differentiation as solution is not unique.

Finite integral

We may be interested in doing integration between only certain values.

$y = \int_a^b 2x dx$

In this case, we use the fundamental theorem of integration which is

$y = \int_a^b f dx = F(b) - F(a)$

$y = \int_1^2 2x dx = x^2 = (2)^2 - (1)^2 = 3$

It represents area under the curve from a to b. If you are driving a tractor on a farm at certain speed(differentiation) then integration from a to b means how much crop did you sow in the area from a to b.

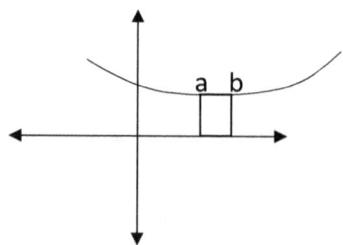

Doing integration is an art. However, the easiest method is-

By substitution

$y = \int \sin(2x + 1) \, dx$

Let $u = 2x + 1$ then

$\frac{du}{dx} = 2$ or $dx = \frac{du}{2}$

$y = \int \sin u \frac{du}{2} = \frac{1}{2} \int \sin u \, du = -\frac{\cos u}{2} = -\frac{\cos(2x+1)}{2} + K$

The constant term is unchanged in an integration.

There are other methods like integration by parts, by fractions and things get pretty complicated. There are integration tables that can be used as well. We will learn advanced techniques as we go long.

Differential equations

The derivatives are mixed up with other functions.

$\frac{dy}{dx} = 5x + 5$

The solution is through integration.

$\int dy = \int (5x + 5) \, dx$

$y = \frac{5}{2} x^2 + 5x + K$

You can check by differentiating the above equation.

The constant of K depends on the initial conditions e.g. if y=2 then x is 4

Will give $2 = \frac{5 \times 16}{2} + 20 + K$

K=-58

$\frac{dy}{dx} - (5x + 5) = 0$

is a first order differential equation.

$\frac{d^2y}{dx^2} - (5x + 5) = 0$

is a second order differential equation.

The second order differential equations are much harder to solve than the first order equations. There are sophisticated techniques to solve the differential equations, which we will learn in subsequent chapters.

Imaginary Numbers

Imaginary numbers are ubiquitous in quantum mechanics. $\sqrt{-1}$ is an imaginary number. The reason is that $(-1)^2$ and $(1)^2$ give 1 as the answer, not -1. So $\sqrt{-1}$ has no real solution. Mathematicians were puzzled when they came across these numbers several centuries ago. But they are now commonly used in mathematics and physics. The symbol for $\sqrt{-1}$ is i, for imaginary. The basic operations are simple e.g. $i^2 = i \times i = -1, i^3 = -i$ etc.

Most people have hard time wrapping their head around imaginary numbers. What do they mean? The name doesn't help either. Let's start with positive real numbers. It is easy to make sense of them. We can count things with them. There are 4 apples, 2 cars, 1 house etc. We can feel and see the positive real numbers.

What about negative real numbers?

Can you touch or feel -4? Obviously if you give away 4 apples, you are left with -4 apples. The negative numbers can show the relationship or transaction of things.

Similarly, zero can explain relative things. I have zero knowledge of quantum mechanics!

What can $\sqrt{-1}$ tell us?

Suppose you have $100 and you plan to loan it to your friend in 10 installments weekly. You will be giving $\sqrt{100}$ = $10 every week as 10× 10=100.

Once you have given away the money, you are left with -$100 in your account book. If you plan to get your money back, in equal installments again then you can calculate $\sqrt{-100}$ which is equal to $10\sqrt{-1}$ or $10\,i$.

You will getting $10\,i$ installments every week and once you get $10\,i$ of them you will have your = $10i \times 10i$ =-100 back. Note everything is imaginary, nothing has happened, you are just hoping that installment will come. They may not! Once you have $10 in your hand, it will no longer be imaginary but real.

The imaginary numbers helped us here to figure out the flow of money but more importantly back and forth phenomenon.

What's the link to quantum mechanics?

First the axis of imaginary numbers is the vertical axis, also called the imaginary or lateral axis.

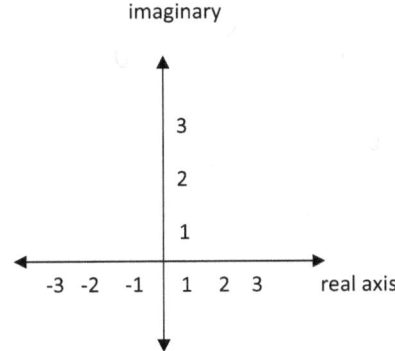

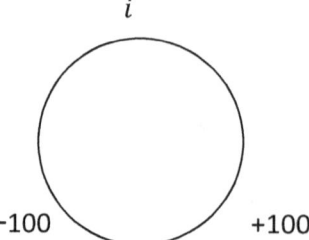

Multiplying i takes 100 to the imaginary axis by rotating it 90 degrees. Further multiplication by i takes it to $100\, i \times i = -100$. You can go back to $+100$ again by multiplying by $i \times i$. As you can see going back and forth is equivalent to moving on a circle as in both cases you go back to the same starting point. The circle here is imaginary.

The quantum mechanics is basically wave mechanics and waves are periodic.

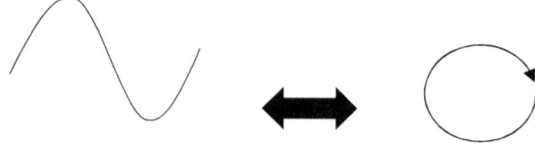

The imaginary numbers encode the periodic information of the waves.

Complex numbers

The real numbers can be mixed with imaginary numbers to make complex numbers.

$2 + 2i$ is a complex number.

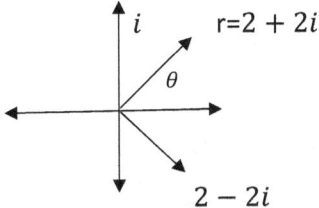

$2 + 2i$ can be represented by a vector. The information contained in the complex number can also be represented by vector r and angle θ.

$2 - 2i$ is the complex conjugate or the shadow of the $2 + 2i$. It is used to get the value of vector r.

$r^2 = (2 + 2i)(2 - 2i) = 4 + 4 = 8$

The relationship between various versions of complex numbers is

$x + iy = r(cos\theta + isin\theta) = re^{i\theta}$

This is the Euler's formula. We can go back and forth between various versions of complex numbers based on convenience.

$e^{i\theta}$ is called the phase, it tells us the starting point of waves.

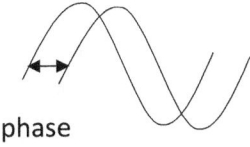
phase

That's enough mathematics for now, let's study some physics.

Chapter 2

Classical Mechanics

Before we dabble with quantum mechanics, we need to review classical mechanics. Even though quantum mechanics is radically different than classical mechanics, many of the mathematical tools and concepts overlap. Classical mechanics is a powerful theory in its own right. Our life revolves around it. It is used in everyday from cars to space shuttles. It pulled the world from dark ages to the modern industrialized economies. For most people classical mechanics is the only physics they know.

Classical mechanics is based on Newton's laws of motion. We have studied them in high school. They are simple, powerful and elegant. F=ma is all you need to get know about the motion of things. The modern industrial revolution is the by - product of the power of classical or Newtonian mechanics. The classical mechanics can get complicated as differential equations can be hard to solve. The physicists and mathematicians over the centuries developed mathematical techniques to solve the equations of Newtonian mechanics. The contributions of great mathematicians and physicists like Euler, Lagrange and Hamilton led to the reformulation of classical mechanics into different branches. There was extensive development of the mathematical structures to solve the equations. It was not until the development of quantum mechanics that physicists realized that the mathematical structures have deep underlying principles of nature buried in them. So, we need to do a brief overview of classical physics.

Hamiltonian Mechanics

Hamiltonian is the total energy of the system. We will develop the equations of motion that tell us the trajectories of particles in terms of the Hamiltonian.

Hamiltonian=KE+PE=$\frac{1}{2}mv^2 + U$ or $\frac{p^2}{2m} + U$

Note K.E.=$\frac{p^2}{2m} = \frac{(mv)^2}{2m} = \frac{1}{2}mv^2$

The coordinates in Hamiltonian mechanics are called canonical coordinates. These are named as p and q. The q is analogous to x or position and p is analogous to classical momentum or p. But they are not restricted to Cartesian coordinates like x, y and z only. The q could represent an angle and p may represent angular momentum etc. The p and q's are called conjugate pairs. The p and q at a specific point could be denoted by $p_i q_i, p_j q_j$ etc.

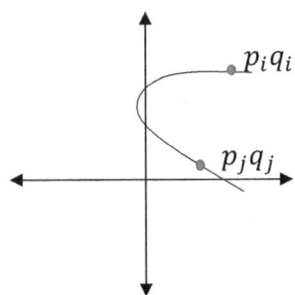

Phase space

Each point on the trajectory has p and q associated with it.

The two fundamental equations of Hamiltonian mechanics are

$\frac{dq}{dt} = \frac{\partial H}{\partial p}$ and $\frac{dp}{dt} = -\frac{\partial H}{\partial q}$

They could be derived from the Lagrangian. We will study it in the next section.

$\frac{dq}{dt} = \frac{dx}{dt} = v$ is the simplest case.

The other side of the equation is $\frac{\partial H}{\partial p} = \frac{1}{\partial p}\left(\frac{p^2}{2m} + U\right) = \frac{2p}{2m} + 0 = \frac{p}{m} = \frac{mv}{m} = v$

The second fundamental equation is

$\frac{dp}{dt} = m\frac{dv}{dt} = ma = F$

The other side is $-\frac{\partial H}{\partial q} = -\frac{1}{\partial q}\left(\frac{p^2}{2m} + U\right) = -\left(0 + \frac{dU}{dx}\right) = -\frac{dU}{dx}$

Is $-\frac{dU}{dx}$ equal to force?

Yes, it is. Because work done or potential energy $U = -\int F dx$ or $F = -\frac{dU}{dx}$

The first equation is velocity written in terms of the Hamiltonian and the second equation is force in terms of the Hamiltonian.

$\frac{dv}{dt} = a$, substituting the Hamiltonian version, we get

$\frac{d}{dt}\left(\frac{\partial H}{\partial p}\right) = a$

$F = ma = m\frac{d}{dt}\left(\frac{\partial H}{\partial p}\right)$

$-\frac{\partial H}{\partial q}$ is also equal to force, so

$m\frac{d}{dt}\left(\frac{\partial H}{\partial p}\right) = -\frac{\partial H}{\partial q}$

$$m\frac{d}{dt}\left(\frac{\partial H}{\partial p}\right) + \frac{\partial H}{\partial q} = 0$$

This is the equation of motion and it helps us know the trajectory of the particle. This version is helpful in solving certain equations that are more difficult to solve with F=ma version.

Lagrangian Mechanics

The lagrangian is defined as KE -PE or T-V. The fundamental principle behind the lagrangian is the principle of least action. It is one of the most important principles in physics. The principle is simple and elegant. The nature is thrifty. It keeps difference between KE-PE as low as possible.

Action or $S = \int_0^t L dt$

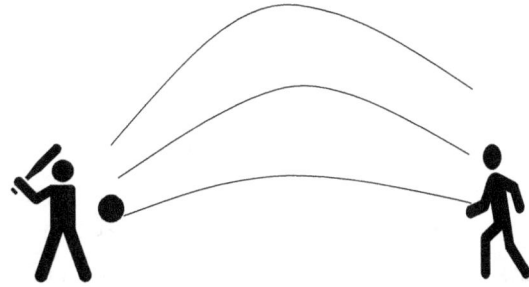

The ball can take several paths to reach the other person at a later time. The path is determined by the principle of least action. The nature choses the path that keeps the KE-PE difference minimum. So, we integrate over all paths and chose the minimum energy difference path. In mathematical terms, we chose the path where action is the least. The technique is part of the calculus of variations.

The equation of motion can be derived from the principle of least action and is called the Euler-Lagrange equation. It determines the trajectory of the ball.

$$\frac{d}{dt}\left(\frac{\partial L}{\partial v}\right) - \frac{\partial L}{\partial x} = 0$$

This equation looks similar to the Hamiltonian equation except the variables are written in terms of the Lagrangian. Here the variables are x and v.

Let's take the simplest case of the trajectory of the ball falling under gravity

$$L = \frac{1}{2}mv^2 - mgx$$

The potential energy is the gravitational energy given by mgx, where x is the height of the ball.

$$\frac{\partial L}{\partial x} = \frac{\partial}{\partial x}\left(\frac{1}{2}mv^2 - mgx\right) = mg = F$$

$$\frac{\partial L}{\partial v} = \frac{\partial}{\partial v}\left(\frac{1}{2}mv^2 - mgx\right) = mv$$

$$\frac{d}{dt}\left(\frac{\partial L}{\partial v}\right) = \frac{d(mv)}{dt} = ma = F$$

Obviously a=-g. The ball falls with acceleration caused by the gravity.

We can choose Hamilton or Lagrange's equations based on convenience. The relative risk/benefit is going too much into classical physics. I will leave at that. What we need to take home is the structure of the classical mechanics as this is what will be used in quantum mechanics. The Lagrangian and Hamiltonian will acquire a deeper meaning in quantum mechanics and will be used in a different way to get the whereabout of the particles.

Conservation Laws

The conservation principles are like guide posts that help particle trajectories to reach their destination. They bring predictability and stability to physics. We are better off with the conservation laws than without them.

What do we mean by conservation?

If a quantity does not change with respect to a variable, we call it conserved.

If a ball on the table is not moving, it is not changing with respect to time.

$$\frac{d(ball)}{dt} = 0$$

The position of the ball(x) is conserved.

If the ball is moving then it will be at a different place at a later time, then

$$\frac{dx}{dt} \neq 0$$

The position is not conserved.

This is known as Noether's theorem.

In the language of Lagrangian, if

$$\frac{\partial L}{\partial q} = 0$$

This means if we change q, it does not affect the Lagrangian L. It also means $\frac{dq}{dt}$ does not affect L.

$$\frac{\partial L}{\frac{\partial q}{\partial t}} = p \text{ is conserved.}$$

The momentum is conserved.

In simpler terms, space translation leads to momentum conservation.

Space translation

Whether the table is on one side of the room or another, it does not affect the movement of ball on the table.

If $\frac{\partial L}{\partial t} = 0$

Then the total energy is conserved.

Taking the example of the ball on the table again. The ball has no kinetic energy, but it has potential energy equal to mgh. The potential energy does not depend on time.

If we measure the height or potential energy of the ball- hourly, daily or weekly, it does not matter.

$$q(t, hourly) \rightarrow q(t, hourly + daily)$$

We can say that time translation leads to energy conservation.

Finally, if $\frac{\partial L}{\partial \theta} = 0$

It is called rotational invariance. The ball on the table will obviously fall on tilting of the table as gravity acts on it. But let's take the example of a fidget spinner and assume that it keeps rotating the same way irrespective of orientation.

$\frac{\partial L}{\frac{\partial \theta}{\partial t}}$ = angular momentum is conserved.

The rotational kinetic energy is equal to $\frac{1}{2}I\omega^2$.

I is the moment of inertia. ω is the angular velocity and is equal to $\frac{\partial \theta}{\partial t}$.

$\frac{\partial L}{\frac{\partial \theta}{\partial t}} = \frac{\partial L}{\partial \omega} = \frac{\partial}{\partial w}(\frac{1}{2}I\omega^2) = I\omega$, which is the definition of angular momentum.

The rotational invariance leads to the conservation of angular momentum.

What's wrong with classical physics?

Life was so much better under classical physics. The physicists were patting themselves at the back. The Newton's laws could explain apple falling from a tree to motion of planets. The electromagnetism laws given by Maxwell explained electricity and magnetism. The thermodynamics told us everything about heat and steam engines. There was complacency among scientists at the end of the 19th century that besides few clouds on the horizon, everything was clear! Some minor unexplained findings were present, but it was felt that it was just a matter of time before the classical physics will solve them as well. They couldn't be more wrong. What unfolded in the next few decades, shook the foundation of physics.

The speed of light is same for every observer irrespective of their frame of reference, was all that Einstein needed to formulate the special theory of relativity. The history of quantum mechanics is not straight forward, there are many people and experiments involved. I will touch on the big events only. It is by no means a chorological or detailed historical account.

The problem of the black body radiation could be considered the starting point of quantum mechanics. A back body absorbs all the radiation aimed at it, gets heated in the process and emits radiation. It is called black body as the radiation emitted at normal temperature is infra-red, which is invisible to the human eye. As it is heated more, it does show color e.g. coal gets red on heating. An ideal black body is a theoretical concept. In practice of course, there are deviations from the ideal black body. The classical physics predicted that the intensity of the radiation depends on the frequency of the radiation. The high frequency radiation carries more energy. Its intensity will keep increasing as we heat the black body. This was called the ultraviolet catastrophe. This was against the experimental results which showed that the intensity peaks and then decreases if we keep on heating the black body.

The classical physics had no answer. In 1900, Max Planck came up with the ad hoc explanation. He came up with the idea that the radiation can only be emitted in small chunks or quanta. Each quanta of energy are related to the frequency by

$E = hf$

He came up with the formula for radiation that was closer to the experimental results, but it lacked sound theoretical basis. But now we know that the quanta are photons. We will study the solution in detail in later chapters.

Let's look at the equation closely. This equation started the quantum revolution. The energy of the quanta depends on how frequently it is oscillating. That's not that revolutionary. If you are jumping up and down, you sure are energetic!

The energy and frequency are related by the constant h, called the Planck's constant. The proportionality constants are common in physics.

$$\frac{E}{f} = \text{constant}$$

Think of measurement units.

$$\frac{1\,foot}{1\,cm} = 30 \cdot 48 \qquad \frac{1\,foot}{1\,inch} = 12$$

The units of measurement and experiments determine the proportionality constant. The value of Planck's constant is $6.6 \times 10^{-34} m^2 kg/s$. The extremely small value of Planck's constant is just a reflection of the units used. We have devised units to use in day to day life like meters, kilogram etc. Obviously, these units are not well suited at the quantum level. You can always choose units for convenience. The Planck's constant is often written as $\hbar = \frac{h}{2\pi}$. This absorbs 2π out of many equations and makes calculations easier. The modified Planck's constant can be set at $\hbar = 1$. The speed of light also can be set at $c = 1$. These are called natural units.

Photoelectric effect

The wave nature of light was well stablished by 19th century. The light splits into colors going into the prism, that's pretty convincing for the wave nature of light. It was realized that shining light on to a metal can eject electrons. The classical physics explanation was that electron absorbs the energy from the light and gets ejected. This means that increasing the intensity of light should cause ejection of higher energy electrons. But the experiments showed that higher frequency light at lower intensity produces more energetic electrons as compared to lower energy light at higher intensity. This goes against the wave nature of light.

The photoelectric effect was successfully explained by Einstein. Even though Einstein became a vocal critic of quantum mechanics later, he was one pioneers of the quantum revolution. It may surprise you that Einstein got Nobel prize not for theory of relativity but for the correct explanation of the photoelectric effect. He took the lead from Max Planck by proposing that the light is made of quanta of energy that collide with the electrons and eject them.

$KE = hf - work\ function$

If the energy of the quanta is more than the work needed to eject the electron, it will happen, otherwise not. The intensity is not a factor.

Take the example of a car wash. If we want to remove the dirt off the car, pouring more water on it is not that helpful but a pressure wash will be more effective. This is what photoelectric effect says.

Double slit experiment

It is customary to include this experiment as it shows the weird nature of quantum mechanics. It is a simple but elegant experiment. It may not have the historical cache that black body radiation has but nevertheless it goes to the heart of the problem in quantum mechanics.

Let's throw a ball at a wall through a slit.

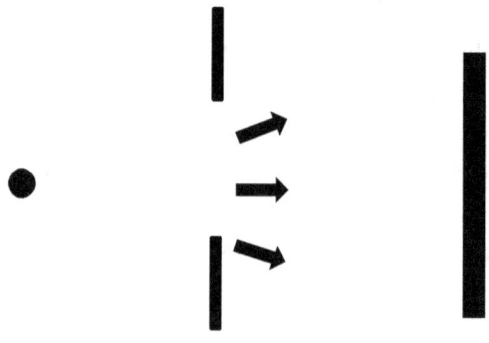

A ball will go through the slit straight through or get deviated by scraping the slits. If we do repeated experiments, the resultant graph will look like a bump in the center that tapers off on either side

What if there are two slits?

We will get two bumps in the results as the ball either goes through one slit or the other.

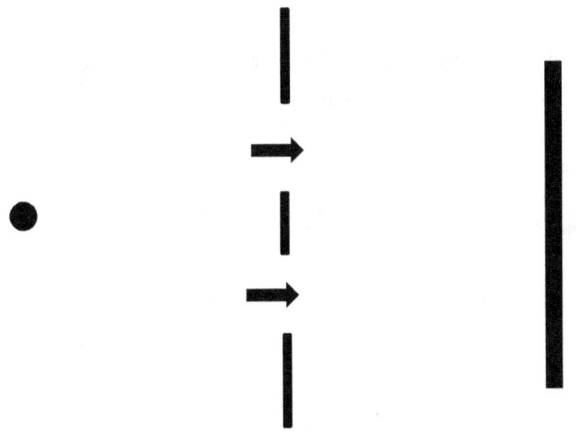

Let's decrease the size of the ball to that of the electron. Does anything change?

If there is one slit, nothing much changes as we still get a bump in the results.

But what about two slits?

This is where we see quantum mechanics reveal itself. We see an interference pattern in the results.

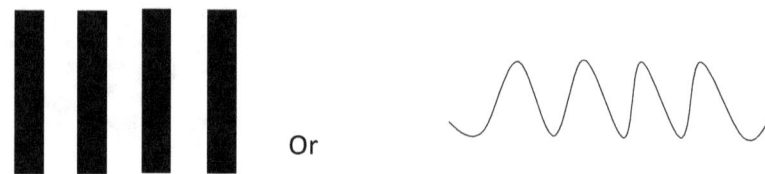

Or

Interference pattern means, electron reaches the plate at certain places but not others. This is not possible for a particle to do. The particle should be able to reach anywhere on the plate. Why would it omit certain places and that too in a pattern form?

The interference is characteristic of waves. The waves can add or subtract each other. This is what makes interference pattern.

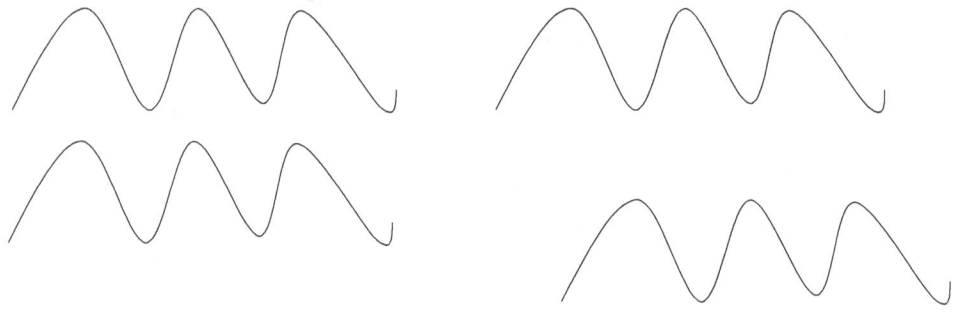

Constructive interference   Destructive interference

As you can see its about matching the pattern. If the waves high and low points do not match, they destroy each other.

But wait a minute, we are taking about a single electron. It is not that water or air is flowing through the slits. How can a single electron interfere with itself?

Which slit the electron went through?

We can try to put a spy camera and see. But nature is clever. If the electron knows that its's being watched, the interference pattern disappears, and we get the classical bumps. This classical quantum mechanics at play. No measurement is gentle. The nature knows it's being watched!

Is electron a particle or a wave?

This of course was the issue with light as well. The wave nature of light was well known but then Einstein showed in photoelectric effect that light is made of small quanta of energy, we now call photons. The light has dual personality, in certain situations it behaves like a wave, in others as a particle. Similarly, electron has dual personality. The electron waves are called matter waves.

The quantum mechanics gives answer to all these questions. But we need to study the laws of quantum mechanics before we answer them. Stay tuned!

Chapter 3

Laws of Quantum Mechanics

The laws of quantum mechanics are quite technical in nature. They can be difficult for the beginner to comprehend. The laws are not straight forward as Newtonian laws of motion. Even Einstein's special theory of relativity is based on simpler principles. It's no wonder that it took many years and required contributions from a lot of physicists to develop quantum mechanics. I am going to take a different approach and try to explain the laws in a more intuitive and natural way than give the comprehensive and didactic version that you see in textbooks.

Everything in quantum mechanics is centered around measurement. Reality is in fact measurement. What we perceive as reality is our senses doing constant measurements by our vision, hearing and smell. The problem with quantum observation is that we cannot use our senses to directly observe quantum phenomenon. We have to rely on mathematics to develop and explain quantum reality. Like Newton's three laws of motion, I will describe three laws of quantum measurement that cover the key concepts of quantum mechanics.

1. Before measurement: If a quantum particle has a choice of states, it occupies all the states at the same time. This is called superposition.
2. Measurement: The measurement leads to the collapse of superposition. The quantum particle chooses one state only.

3. After measurement: The chosen state could be any one of the states available. The result is completely random.

If you think about these rules and apply to our daily lives, it will distort our reality.

What do we mean by a quantum state?

A quantum state may refer to position, momentum, energy, spin etc. If a particle has a choice of being at different places, then it occupies all the available spaces at the same time.

The act of measurement is not a mere observation. There is nothing like a gentle observation. The measurement always leads to the disturbance of the quantum state. It leads to the collapse of the superposition. How does measurement leads to the collapse of the superposition is a mystery to this day.

Does this mean that particle has no fixed position before the measurement?

It seems so. The reality of a particle at a position is created by our act of watching it. If we do not observe the particle, it has no reality or fixed position. This was hard to swallow for some physicists including Einstein. Do Sun and planets need humans to observe them to make them real?

The result of the measurement is random which means any of the available choices can be the result of the measurement.

This is how bizarre the quantum rules are. If you want to know where I am, I could be anywhere in the universe. I occupy all the available positions in space at the same time. Once you decide to look at me, I will randomly appear anywhere in the universe.

When we develop the mathematics of quantum mechanics, there are restrictions put on quantum states so that some bizarre results are avoided. The process of normalization of the wave function avoids the results of the finding particle at the ends of the universe.

Even physicists who developed quantum mechanics were aware of the bizarre nature of the quantum laws. I am sure you must have heard about the Schrodinger's cat.

If we put cat in a box and there is a poison gas that can be released by random radioactive decay based on quantum laws, then before we open the box

$Cat = Dead + Alive$

The cat is in superposition of dead and alive at the same time. When we open the box, the act of measurement leads to the collapse of the superposition. The result is random which means there is 50:50 chance of getting a dead or an alive cat.

This shows how absurd the quantum laws are if applied directly to classical objects like us. But experiment after experiment has shown the remarkable accuracy of the quantum mechanics. As we go from smaller to bigger size e.g. from electron to atom to molecule, the quantum phenomenon gets diminished and classical physics takes over. If we take average of the results predicted by quantum mechanics, we get something that resembles classical physics. When does turn over from quantum mechanics to classical physics takes place, is a subject of research.

The laws that I described earlier are based the orthodox interpretation. There was always controversy around the interpretation of quantum mechanics. With the laws like these, it is no surprise. During Solvay conference in 1927, the founders of quantum mechanics came up with the orthodox interpretation. It was later called Copenhagen interpretation. It believes in the random and probabilistic nature of quantum mechanics. This interpretation is widely accepted and backed by lot of evidence. Not everyone was convinced, the most famous dissenting voice being that of Einstein. He believed that quantum mechanics is an incomplete theory. It is the short coming of the theory when it does not know the position of the particle and the formation of super position is due to lack of information. There are hidden variables that determine the position of the particle and there is no randomness involved. Niels Bohr was a strong proponent of the orthodox interpretation. Many years later, John Bell came with the Bell's theorem which gave convincing evidence that quantum mechanics is indeed random and probabilistic theory. We will explore it in detail later.

For history buffs, search for the group picture taken at the Solvay conference. It a rare event in history where such brilliant physicists were assembled at one place, Einstein, Bohr, Marie Curie, Planck, Pauli, Dirac, Heisenberg, Schrodinger to name a few!

The rest of the chapter is devoted to developing the mathematics needed to describe these laws and how to apply them.

How do we describe a particle?

The particle could be an electron, proton, neutron etc. The thing to remember is that all particles are waves at heart. These are called matter waves. De Broglie gave this revolutionary idea that there are waves associated with each particle

$$\lambda = \frac{\hbar}{mv}$$

The wave length for classical object is so small that its virtually undetectable. This is due to the extremely small value of the Planck constant, $h = 6 \cdot 62 \times 10^{-34} m^2 kg/sec$. But quantum particles are small and move at high speed, so wave length is detectable.

Waves by nature are spread out. The waves can interfere with each other causing constructive and destructive interference. The wave like properties can explain a lot of weird quantum phenomenon. Remember the matter waves are very different than classical waves. We see ocean waves or wind, but these waves are no quantum waves. The classical waves are ordinary particles moving in a periodic fashion to create a wave. The particle themselves are not waves. The particles need a medium to travel. The quantum or matter wave is very different. Even a single particle has wave properties. The waves of a single particle can spread and interfere with itself without involving any other particle.

We need to select a symbol for a quantum particle. The traditional symbol is the Greek letter phi, Ψ. It is appropriately called the wave function. It is a mathematical function that tells us everything about the particle like its position, energy, momentum etc. The waves are periodic phenomenon. To describe a periodic phenomenon, we need to use complex numbers. So, wave function is a complex number in most cases.

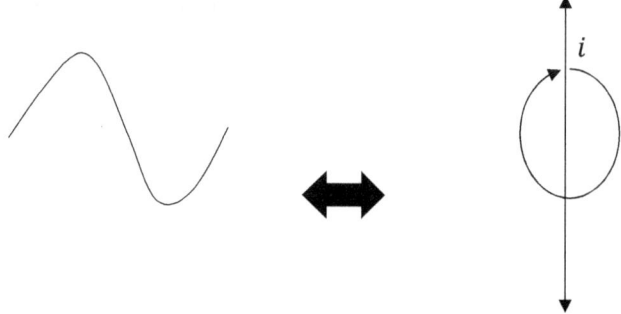

A wave completing a cycle is equivalent to a circle on the imaginary axis. The complex numbers have real and imaginary part which encode the periodic motion. The result of experiments in the end has to be a real number. How would you interpret $2 + 2i$?

To get rid of the complex number in the end result, we need to create the complex conjugate of the wave function. The complex conjugate of $2 + 2i$ is $2 - 2i$. The symbol of complex conjugate of the wave function is Ψ^*.

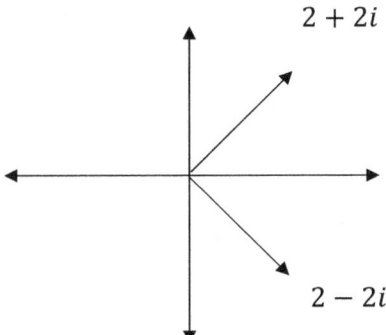

Think of complex conjugate as a mirror image of the wave function. We are not complete without our shadow, so wave function needs the complex conjugate to complete it.

The probability of finding a particle between two points a and b is given by

$\int \Psi\Psi^* dx$ or $|\Psi|^2 dx$

This is called Born interpretation. The wave function is also sometimes called the probability amplitude. Squaring it gives the probability of finding the particle.

In the above case, probability density is $(2 + 2i)(2 - 2i) = 8$.

The wave function has to be reasonable. You cannot cook up any wave function. Reasonable means if you are looking for a particle in your room, wave function has to find the particle in a room. It should have a higher probability of finding the particle in a room and it should decrease outside the room.

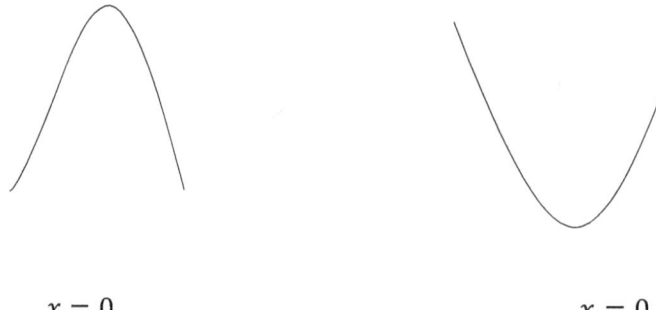

x = 0   x = 0

The wave function on the left is a reasonable wave function. It has the higher probability where you expect the particle to be and tapers off as we move away. The technical name of the reasonable test is called normalization. The parabolic wave function is not a reasonable wave function. It predicts that particle would rather be at the end of the universe than in the room!

The condition for normalization is

$$\int_{-\infty}^{+\infty} |\Psi|^2 dx = 1$$

This equation simply states that particle has to be found somewhere.

Superposition

Each point in space is part of some wave function. There are different wave functions for finding the particle in the center of the room, outside the room, on the floor, on the roof etc.

$$\Psi = c_1\psi_1 + c_2\psi_2 + c_1\psi_2.... = \Sigma c_n\psi_n$$

The coefficients are the probabilities of finding each wave function or the particle at certain position.

e.g. $\frac{1}{\sqrt{2}}\psi_1 + \frac{1}{\sqrt{2}}\psi_2$. Here $c_1 = c_2 = \frac{1}{\sqrt{2}}$.

The probability of finding the particle at ψ_1 and ψ_2 = $|c_1|^2 = |c_2|^2 = \frac{1}{2}$

There is 50 % probability of finding the particle at position 1 and position 2. The states are normalized as total probability is $\frac{1}{2} + \frac{1}{2} = 1$.

These Ψ's need a place to live. We count our money in real numbers and the collection of all real numbers is the real number space. The wave functions are complex functions and they are vectors that live in the Hilbert's space. The vectors span the space which means nothing is left empty, all of the space can be described by the quantum state vectors. The complex conjugates live in the dual space.

New notation

The modern quantum mechanics uses the Dirac notation to describe the wave functions. It is simpler to use as you will see.

We need to put the wave function behind bars! $|\Psi\rangle$ is called a ket vector.

The complex conjugate wave function Ψ^* is imprisoned as well. $\langle\Psi|$ is called the bra vector. In this notation, no need to put $*$ on the wave function.

This notation is also called bra-ket notation.

Usual mathematical operations can be done on these vectors

$|\Psi\rangle + |0\rangle = |\Psi\rangle$

$|\Psi\rangle + |-\Psi\rangle = |0\rangle$

$5 \times \Psi = 5\Psi$

Similar operations can be done on bra vectors

$\langle\Psi| + \langle 0| = \langle\Psi|$

Inner product

It is analogous to the dot product in classical physics.

$a.b = ab\cos\theta$

It projects b onto a. If you pull a box at an angle towards you then the horizontal force on the box is given by the dot product. It calculates the horizontal component of the force.

The dot product in quantum mechanics is called the inner product.

$\langle a|b \rangle$

The two vectors do not have to be complex conjugates of one another, so you can mix and match any vector. The result is a complex number. The probability is given by squaring it to convert the complex number into a real number answer.

$P = |\langle a|b \rangle|^2$

$\langle a|a \rangle = 1$

If you project a vector onto itself, you get the same thing. Note cos 0=1.

If two vectors are perpendicular, then

$\langle a|b \rangle = 0$. Here cos 90 =0.

Here b is not projecting anything in a direction. These are called orthogonal vectors. If they are normalized, then they are called orthonormal vectors.

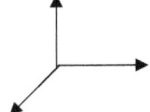

The orthonormal vectors are good candidates for basis vectors. The x, y and z directions are examples of orthonormal vectors.

I do not want to get too technical. If two states are clearly different, they are orthonormal e.g. head and tails of a coin can form orthonormal basis for the coin.

There is a short hand to write the property of orthonormal vectors.

$$\langle a|b\rangle = \delta_{a,b} \begin{Bmatrix} 1, a = b \\ 0, a \neq b \end{Bmatrix}$$

This is called delta function. You will see it a lot in quantum mechanics. It deserves a formal introduction.

Kronecker Delta

$$\delta_{i,j} \begin{Bmatrix} 1, i = j \\ 0, i \neq j \end{Bmatrix}$$

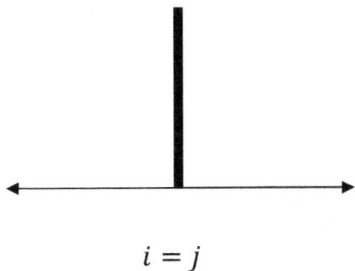

$i = j$

It is a simple concept. The function is peaked at one point only, everywhere else its zero.

Say I am in Toronto, let's create a delta function.

$\delta_{i,j}$ i=Toronto, j=another city

If you look for me in different cities, you will find

$\sum_{i=1}^{3} c_i \delta_{i,j}$, here c=city and you look for me in three cities ranging from

1= Tornoto, 2= Vancouver, 3= Montreal

$= T \times 1 + V \times 0 + M \times 0 = Toronto$

The function is only peaked at Toronto, everywhere else its zero. Sticking in δ function collapses the sum to a single point.

Dirac Delta function

It is a variation of the Kronecker delta and is useful for continuous variables. Here the sum is replaced by integral. You can say this function is custom made by Dirac to work in quantum mechanics.

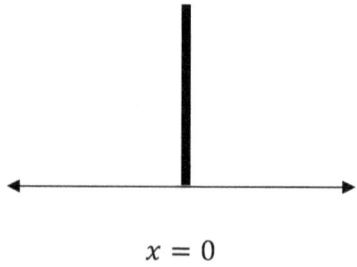

$x = 0$

$$\delta(x) \begin{Bmatrix} \infty, x = 0 \\ 0, x \neq 0 \end{Bmatrix}$$

If x= my location and you have to search me all over Canada, then

$$\int_c f(x)\delta(x)dx = f(x)$$

The delta function picks my location out of the integral.

The peak in the function may not be at x=0. Say it's some distance away from the origin, then the delta function is $\delta(x - a)$

$$\int_c f(x)\delta(x - a)dx = f(a)$$

It just gives us a freedom to put the particle anywhere, not just at x=0.

Matrix notation

I know you are getting sick of notation, but matrix notation is unavoidable. It is very convenient and gives useful insight into the calculations. Historically, matrix formulation was developed independently at the same time as the wavefunction formulation. Matrix formulation was developed by Heisenberg. We won't go into the full matrix formulation as its unnecessary. In modern quantum mechanics, wave function, Dirac notation and matrix formulation are used interchangeably based on convenience.

The state vector can have different components. If one is moving, there are components in x, y, and z directions. The quantum vectors are not restricted to the spatial directions though.

$$a = c_1 a_1 + c_2 a_2 + c_3 a_3$$

$$b = n_1 b_1 + n_2 b_2 + n_3 b_3$$

In matrix form

$$|b\rangle = \begin{pmatrix} b_1 \\ b_2 \\ b_3 \end{pmatrix}$$

The ket vector can be written as a column vector.

$\langle a| = (a_1^* \quad a_2^* \quad a_3^*)$

The bra vector can be written as a row vector.

The inner product $\langle a|b \rangle$ is converted into multiplication of row and column vectors.

$$(a_1^* \quad a_2^* \quad a_3^*) \begin{pmatrix} b_1 \\ b_2 \\ b_3 \end{pmatrix}$$

$= a_1^* b_1 + a_2^* b_2 + a_3^* b_3$

I left out the coefficients(c,n) to keep equations cleaner but obviously you have to multiply them as well.

Before we go on to more mathematics, let's do a recap and take the example of a coin flip to demonstrate the quantum laws of measurement.

1. Before measurement- The quantum coin state has two choices, heads or tails. So, it occupies both of them and forms a superposition.

$\Psi = \frac{1}{\sqrt{2}} \Psi_H + \frac{1}{\sqrt{2}} \Psi_T$ or $\Psi = \sum c_n \Psi_n$, n is either heads or tails, $c_n = \frac{1}{\sqrt{2}}$

This is the wave function formulation.

The equivalent Dirac notation is

$|\Psi\rangle = \frac{1}{\sqrt{2}} |\Psi_H\rangle + \frac{1}{\sqrt{2}} |\Psi_T\rangle$

The matrix formulation looks like

$$|\Psi\rangle = \frac{1}{\sqrt{2}}\begin{pmatrix}1\\0\end{pmatrix} + \frac{1}{\sqrt{2}}\begin{pmatrix}0\\1\end{pmatrix}$$

Here we have arbitrarily chosen heads to be 1 and tails to be 0.

2. Measurement- Superposition collapses to a single state, either heads or tails.

3. After measurement- The result is completely random. We cannot predict if heads or tails will be the result. But we can tell the probability of getting heads or tails.

The probability of getting heads is $|c_H|^2 = |\langle H|\Psi\rangle|^2 = \left(\frac{1}{\sqrt{2}}\right)^2 = \frac{1}{2}$

The probability of getting tails is $|c_T|^2 = |\langle T|\Psi\rangle|^2 = \left(\frac{1}{\sqrt{2}}\right)^2 = \frac{1}{2}$

Total probability adds up to 1.

The heads and tails are orthonormal states

$\langle H|T\rangle = (1\;0)\begin{pmatrix}0\\1\end{pmatrix} = 1 \times 0 + 0 \times 1 = 0$

I am sure you have many questions in your mind. Let me ask them on your behalf.

How the hell do we know what happens before the measurement?

This is not a trivial question. This question was asked by none other than Einstein. He argued that the need for superposition is due to lack of information. The quantum theory is incomplete. There are hidden variables that determine if the coin is heads or tails before the measurement. The orthodox interpretation says that coin has no fixed state, before measurement. Only on measuring, the coin decides to be either heads or tails. Quantum mechanics is a probabilistic theory. It means we need to gather lot of data. The data and the results fit the picture of superposition perfectly.

What constitutes a measurement?

We know that measurement act is not just being an observer. The measurement causes collapse of the wave function. How the collapse of the superposition happens is unknown.

Does it always involve a conscious living being like a human? Can a cat or dog observing a quantum system cause the collapse of the superposition? What about non-living things? Can macroscopic bodies like sun or stars cause the collapse of the wave function?

The orthodox interpretation is that these questions are not scientific. There is no need to answer philosophical questions. The non-living things have no concept of observation. Cats and dogs cannot communicate quantum observation to us!

How do we know results are random?

This is obvious, do lots of experiments. If you do a single experiment and find a particle somewhere, you just assume it's there. You have to collect millions of particles that are in the same state e.g. ground state and then correlate results. The measurements have to be done on separate particles in the same state. Once you get a result on a particle, it will remain the same, even if you check it 10 times. The reason is once you have disturbed the state, it is not going to go to the superposition again. Each time you find the position, mark it on a graph.

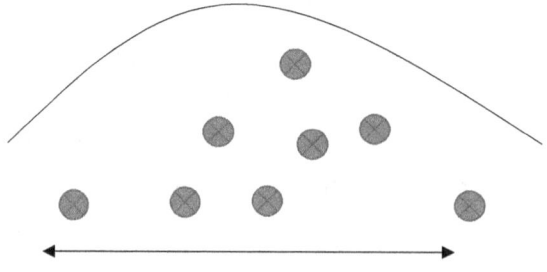

After doing millions of experiments, the plots can look like a graph. The best you can tell is that the result will be somewhere in the graph with high probability at the center, but the individual result is still random, it can be anywhere in the graph.

You may say that it happens in classical flipping of coin too. We all know that there is 50:50 chance of getting heads or tails. But if you do 10 flips, the result may be 6 heads and 4 tails. This does not disapprove the probability. You have to flip millions of coins to come close to the 50:50 probability. You may think that the coin flip is random for an individual result. But this is not really true. It is theoretically possible for a person to learn the trick to throw the coin in such a way to get a particular result. This does not work in quantum mechanics. There is no way to secure a desired result.

Schrodinger Equation

I hope you are getting the idea of wave function, superposition and random results. The purpose of physics or science for that matter is to predict things. The equations are written so that we can forecast events. It is true that an individual quantum event will be random but at least we can give it a probability. This means we need to solve the wave function. Solving wave function requires an equation. F=ma is the equation that governs classical mechanics. Similarly, the Schrodinger equation is the fundamental equation of quantum mechanics. It is not directly derived from anything but take it as a law. There are some clever techniques to derive it but for our purpose taking it as a fundamental equation is reasonable and everything else flows from it.

$$i\hbar \frac{\partial \Psi}{\partial t} = H\Psi$$

H is the Hamiltonian or total energy of the system.

What does this equation tell us?

We will explore the mathematical details but at a basic level it tells us that how a particle changes with time, depends on its energy. This is obvious. If I am sitting on a chair and have no kinetic and potential energy, then at a later time I will still be sitting on the chair. If I have some kinetic energy and potential energy, then I can climb stairs and you will find me at a different place at a later time. Things are not that simple in quantum mechanics. First of all, Ψ is a mathematical function which is a complex number. The complex conjugate of Ψ also has to be considered. We will explore what it means by time evolution of the wave function.

What about $i, \hbar$ in the equation?

There is no avoiding them in quantum mechanics. The complex number i encodes the periodic nature of the waves and the Planck constant $\hbar$ relates energy of the wave to how fast its oscillating.

Operators

Before we solve the Schrodinger equation, we need to know about operators.

The operators are functions that act on the wave function. Even in classical physics there are operators. The velocity of an object is $\frac{dx}{dt}$. The operator in this case is $\frac{d}{dt}$, which tells us how much distance is cover in certain time.

In quantum mechanics, operations are clearly shown as they act on the wave function. There are special operators that act on the wave function and give experimental results. These are called Hermitian operators.

Think of Hermitian operators as measuring devices. They act on the wave function, which we need to measure and give experimental results.

Operator × wave function= experimental result

The important property of Hermitian operators is

$$\langle B|\hat{H}|A\rangle = \langle A|\hat{H}|B\rangle^*$$

$\langle H \rangle = \langle H \rangle^*$

The hat on the top of H indicates it is an operator. Some books faithfully use this notation, others ignore it. It is also customary to write the complex conjugate of the operator with the dagger symbol, $H^\dagger$.

The above equation says that when the operator acts on ket vector A then on the bra vector B, it is equivalent to reversing the order and taking the complex conjugate of it.

The intuition behind it is simple. The operators have real values. $\langle H \rangle$ is symbol for average value of all results. It is a real number. The real number is its own complex conjugate e.g. $5^* = 5$. Think of this way, Hermitian operator measures your speed, the complex conjugate operator measures the speed of your shadow. Obviously, you and your shadow move at the same speed!

The Hermitian operators are also called observables. The position, momentum and energy operators are Hermitian, their results are real.

The expectation value comes over and over again. It is important to learn about it.

Take the case of dice. We can get 1,2,3,4,5 and 6 if we throw a dice. Each value has $\frac{1}{6}$th probability of occurring.

The average or expectation value of dice is

$1 \times \frac{1}{6} + 2 \times \frac{1}{6} + 3 \times \frac{1}{6} + 4 \times \frac{1}{6} + 5 \times \frac{1}{6} + 6 \times \frac{1}{6} = \frac{1}{6}(1 + 2 + 3 + 4 + 5 + 6)$

or $\frac{21}{6} = 3 \cdot 5$

In the language of quantum mechanics, the dice wave function can be written as

$|\Psi\rangle = c_1\Psi_1 + c_2\Psi_2 + c_3\Psi_3 + c_4\Psi_4 + c_5\Psi_5 + c_6\Psi_6$

The probability coefficient is

$|c_1|^2 = |\langle\Psi_1|\Psi\rangle|^2 = \left(\frac{1}{\sqrt{6}}\right)^2 = \frac{1}{6}$

Similarly, all the other coefficients can be calculated, and they are obviously same.

Each value of dice ranges from 1 to 6, let's call it λ.

$\langle D \rangle = \text{average} = \sum_{i=1}^{6} \lambda_i |c_i|^2 = \sum_{i=1}^{6} \lambda_i |\langle\Psi_i|\Psi\rangle|^2$

This is what we did earlier to get the value of 3.5.

We have to prove that $\langle D \rangle = \langle \Psi | \hat{D} | \Psi \rangle$

$|\Psi\rangle = c_1\Psi_1 + c_2\Psi_2 \ldots\ldots$

$\langle\Psi| = c_1^*\Psi_1^* + c_1^*\Psi_1^* \ldots$

When the operator D acts on $\Psi\rangle$, we get each value λ

$D\Psi_1 = \lambda_1 \Psi_1$

We will formally introduce this equation later.

$D|\Psi\rangle = \lambda_1 c_1 \Psi_1 + \lambda_2 c_2 \Psi_2 \ldots\ldots$

Then we multiply by $\langle\Psi| = c_1^*\Psi_1^* + c_1^*\Psi_1^*$

We will get terms like

$|c_1|^2 \lambda_1 \langle\Psi_1|\Psi_1\rangle \ldots + |c_1|^2 \lambda_1 \langle\Psi_1|\Psi_2\rangle$

The states are orthonormal so $\langle\Psi_1|\Psi_1\rangle = 1$ and $\langle\Psi_1|\Psi_2\rangle = 0$

We will be left with

$|c_1|^2 \lambda_1 + |c_2|^2 \lambda_2 .. = \sum_{i=1}^{6} \lambda_i |c_i|^2 = \langle D \rangle$

The Hermitian operators can be written in a matrix form.

There is a quick way of checking if an operator is Hermitian. The diagonal elements are real and off diagonal elements are complex conjugates of each other.

$H = \begin{pmatrix} 1 & 1+i \\ 1-i & 2 \end{pmatrix}$

The diagonal elements are real 1 and 2 and off diagonal elements are complex conjugates.

$\begin{pmatrix} i & 1+i \\ 1-i & 4+i \end{pmatrix}$ is not Hermitian.

The Hermitian operator for position is x. That's simple.

The Hermitian operator for momentum is $-i\hbar \frac{\partial}{\partial x}$.

In case you are not convinced that these are the Hermitian operators, you are welcome to test it by using property $\langle B | \hat{P} | A \rangle = \langle A | \hat{P} | B \rangle^*$ .I won't prove it here, take it for granted.

The operator for energy is the Hamiltonian. We can derive it from the momentum operator. The Hamiltonian or total energy is K.E.+P.E.

$H = \frac{1}{2}mv^2 + U = \frac{p^2}{2m} + U$ where p=mv

Putting in the operator for p, $-i\hbar \frac{\partial}{\partial x}$

$H = \frac{-\hbar^2}{2m} \frac{\partial^2}{\partial x^2} + U$

Once we have the Hamiltonian in operator form, we are ready to solve the Schrodinger equation

$$i\hbar \frac{\partial \Psi}{\partial t} = H\Psi$$

$$i\hbar \frac{\partial \Psi}{\partial t} = (\frac{-\hbar^2}{2m} \frac{\partial^2}{\partial x^2} + U)\Psi$$

Separating distance and time can simplify the equation

$$\Psi(x,t) = \psi(x), \phi(t)$$

$$\frac{\partial \Psi}{\partial t} = \frac{d\phi}{dt}\psi \quad (\psi \text{ does not depend on t, we did partial product differentiation here})$$

$$\frac{\partial^2 \Psi}{\partial x^2} = \frac{d^2\psi}{dx^2}\phi \quad (\phi \text{ does not depend on t, we did partial product differentiation here})$$

Substituting these values in the Schrodinger equation

$$i\hbar \frac{d\phi}{dt}\psi = \frac{-\hbar^2}{2m} \frac{d^2\psi}{dx^2}\phi + U\psi\phi$$

Further separating ψ and ϕ

$$i\hbar \frac{1}{\phi}\frac{d\phi}{dt} = \frac{-\hbar^2}{2m} \frac{d^2\psi}{dx^2}\frac{1}{\psi} + U$$

We have time dependence on one side and distance dependence on another, but they are equal to each other. This means each term is equal to some constant.

Why is that?

Logic is simple.

Every Wednesday (time dependence) =I go somewhere (distance dependence)

Each equation is equal to a constant. Can you guess?

Gym!

We will call the constant E for the Schrodinger equation.

$$i\hbar \frac{1}{\phi}\frac{d\phi}{dt} = E$$

This equation is not too hard to solve.

$$\frac{d\phi}{dt} = \frac{E\phi}{i\hbar} = \frac{-iE}{\hbar}\phi$$

Whenever you get the same function back after differentiation, the answer is exponential function.

$$\phi = e^{\frac{-iE}{\hbar}t}$$

Now the other part of the equation

$$\frac{-\hbar^2}{2m}\frac{d^2\psi}{dx^2}\frac{1}{\psi} + U = E$$

$$\frac{-\hbar^2}{2m}\frac{d^2\psi}{dx^2} + U\psi = E\psi$$

This is the *time independent* Schrodinger equation.

In short $H\psi = E\psi$

The solution of this equation will depend on the choice of Hamiltonian.

Solving equation is one thing, understanding is another. Let's get the intuition behind the equations.

First let's take the case of the time dependence factor.

$$\phi = e^{\frac{-iE}{\hbar}t}$$

$$\Psi(x,t) = \psi(x)e^{\frac{-iE}{\hbar}t}$$

The time dependence factor is called the phase factor.

What's a phase?

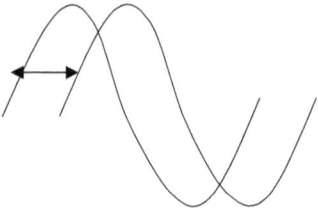

The difference in the starting point between two waves is called the phase. We are solving the time dependence of the wave function. This phase can represent the movement of wave with time. But the wave function is a mathematical function that tells us the probability of finding a particle. Does this phase influences it?

It turns out that it does not. The starting point is arbitrary. The waves are periodic, so things keep on repeating. The starting point does not matter.

Think of this way, the second hand of the clock keeps on repeating itself.

What's the difference in going from 10 seconds to 70 seconds vs 20 seconds to 80 seconds?

From the perspective of the clock, none. The stating point is arbitrary, the second clock completes a circle and reaches where it starts. So, the phases or the starting points have no relevance for the second clock.

But there are things that depend on time. We lead over lives based on time. The choices of time are obviously arbitrary, Eastern time, Pacific time etc. But our daily schedule is impacted by them.

Similarly, the probability of the wave function is not impacted by the time phase. However, there are detectable differences that phases can cause e.g. Berry phase. This is too sophisticated for us at this stage, I will explain it in later chapters.

How does the phase factor cancel out in probability calculation?

The mathematics is designed in such a wave that if the phase factor is $e^{-i\theta}$ then the complex conjugate of the phase is $e^{+i\theta}$.

The probability is $\Psi^*\Psi = \psi(x)e^{\frac{-iE}{\hbar}t}\psi^*(x)e^{\frac{+iE}{\hbar}t} = \psi\psi^*$

The phase factor cancels out and the probability is time independent.

It should not be surprising. The shadow has to move in tandem with the object, there cannot be any phase between them when calculating where the object is.

What to make of the time independent Schrodinger equation?

$H\psi = E\psi$

H is an observable. It acts on the wave function to give E as the result of the observation. E is the energy value.

There is a better way of understanding it. Think of operators as measuring devices. They measure the wave function. The results are real and experimentally verifiable.

Measuring device × Object to be measured=Measurement result × Object

How do we check our weight?

$W \, \Psi_{weight(60)} = 60 \, \Psi_{weight(60)}$

I stood on the weighing scale and it showed 60kg (when I was younger!)

In the language of quantum mechanics, the weighing scale operator acts on my weight function and results in the value of 60kg. The results are called eigenvalues. The time independent Schrodinger equation is also called the eigenvalue equation.

If I am 70kg then my weight function is different. The equation will read

$$W \Psi_{weight(70)} = 70 \Psi_{weight(70)}$$

There are lot of weight functions that the weighing scale operator acts on, giving different eigenvalues. The weight functions are collectively called eigenfunctions. Each eigenfunction has a probability of occurring, $c_w \Psi_w$. The probability is given by $|c_w|^2$.

The probability will depend on experimental results. For an average young man, you expect the probability of 70kg will be high and of 30kg, much lower.

Which eigenfunctions Ψ_{weight} are included and their nature depends on solving the Schrodinger equation. It depends on the weighing scale operator. The operator may be designed in such a way that it cannot tell difference of less than 1kg. It may not be able to weigh more than 200kg. These characteristics will determine the eigenfunctions. $\Psi_{weight(220)}$ is not an eigenfunction of such a weighing scale operator.

The collection of the eigenfunctions along with their corresponding values and the probability coefficients form the whole solution of the Schrodinger equation.

$$\Psi_{weight} = \sum_{n=1}^{200} c_w \Psi_w$$

Different operators may or may not share eigenfunctions.

$$T \Psi_{tall(6)} = 6 \Psi_{tall(6)}$$

If T is the tallness operator that measures the height. So, it can measure a person to be 6 feet in height. It is possible that a 6 feet person also weighs 70kg. In that case the operators W and T share eigenfunctions.

$Ha\, \Psi_{nail} = 3\, \Psi_{nail}$

We can imagine a Hammer operator that acts on a nail and inserts it 3mm in the wall. Obviously Ψ_{nail} and Ψ_{weight} are entirely different. In this case, W and Ha operators do not share eigenfunctions.

The advantage of sharing eigenfunctions is that operators can be measured at the same time. In the above case, weight and height can be measured at the same time.

Don't you think it is mathematics run amok! What's the point of this tedious way of measuring things?

It is a valid point for our daily lives. The above mathematical formulation of measuring things creates unnecessary complexity without adding anything.

But it is very useful in quantum mechanics. It allows superposition to be created. If I am a quantum particle, I have no fixed weight before the measurement. I can have any weight from 1 to 200. I am in superposition of all weights. Once I get on a weighing scale and someone reads the results, my superposition collapses. The result that comes out is random. You could see 1kg or 200kg. The probability of each value will only be known after doing repeated experiments and counting how many times each value comes out. If this is not weird, I don't know what else is!

In summary Hamiltonian acting on the wave function yields energy eigenvalues. The energy eigenfunctions range from the ground state(n=1) to higher energy functions.

$\Psi = \sum_{n=1}^{\infty} c_n \Psi_n$

Where's time in all this?

We already learned that time introduces a phase that cancels out when calculating probability.

$$\Psi(x,t) = \sum_{n=1}^{\infty} c_n \Psi_n \, e^{\frac{-iE_n}{\hbar}t}$$

In the language of operators, time is described in terms of unitary operator. It is like multiplying by 1. It does nothing to the magnitude of the function.

Time operator=U

Complex conjugate= U

U U†=1

Actually 1 in operator language is called I, the identity operator.

I in matrix form looks like $\begin{pmatrix} 1 & 0 \\ 0 & 1 \end{pmatrix}$

The time operator U is not a Hermitian operator as its eigenvalues are of the form $e^{-i\theta}$, which are not real.

$$U|\Psi_0\rangle \rightarrow |\Psi_t\rangle$$

The purpose of the time in this form is to preserve distinction of states as they evolve in time. An electron remains an electron, a proton remains a proton after their evolution in time.

Does it mean nothing changes with time?

That's not what we see in life, things change all the time.

The time does its magic when Hamiltonian is time dependent. It is not hard to understand. If the potential is time dependent, it's going to impact particles. If you switch magnetic field off and on, particle trajectories are going to be impacted.

How do we describe time dependence of the Hamiltonian?

$$\Psi = c_1 \Psi_1 + c_2 \Psi_2$$

The different eigen functions have certain probability. The time dependence of the Hamiltonian causes the probability coefficients to become time dependent.

$c_1(t)$

This means the probability of getting a particular result will change with time if Hamiltonian is time dependent.

Let's go to the world of politics to see how it works.

It's fair to say that once a politician, always a politician. The political ideologies are stationary. A socialist or a conservative have their thinking frozen in time. So, we can assume they are not time dependent.

$$\Psi = c_1 \Psi_D + c_2 \Psi_R$$

Our political wavefunction is made up of democrats and republicans. The operator is, say midterm elections. If nothing in the election operator changes, both parties have equal probability of winning.

$$\Psi = \frac{1}{\sqrt{2}} \Psi_D + \frac{1}{\sqrt{2}} \Psi_R$$

$$|c|^2 = \frac{1}{2}$$

The eigen value equations will be

$H_{election} \Psi_D$ = Democrat win Ψ_D

$H_{election} \Psi_R$ = Republican win Ψ_R

If $H_{election}$ is time dependent, which means there are political winds blowing that change the election race. A blue wave will cause the probability of democrat win to increase and a red wave is going to boost the republican majority. These waves come alternately in elections as people get fed up with a political party.

In a time - dependent Hamiltonian with blue wave, the probability coefficients can look like

$$\Psi = \frac{2}{\sqrt{3}} \Psi_D + \frac{1}{\sqrt{3}} \Psi_R$$

This is how time causes change in our life. It is a very technical subject in quantum mechanics and will discuss it in detail in the chapter on time dependent perturbation theory.

Free particle

Enough of political talk, let's do some real quantum mechanics and solve the simplest case of the Schrodinger equation. A free particle means it has no potential in the Hamiltonian.

$$H\psi = E\psi$$

$$\frac{-\hbar^2}{2m}\frac{d^2\psi}{dx^2} + 0 = E\psi$$

$$\frac{d^2\psi}{dx^2} = -\frac{2mE}{\hbar^2}\psi$$

We can soak up factors by introducing a new variable

$$k = \frac{\sqrt{2mE}}{\hbar}$$

$$\frac{d^2\psi}{dx^2} = -k^2\psi$$

The solution is exponentials. It is a second order differential equation, so there are two solutions.

$$\psi = c_1 e^{ikx} + c_2 e^{-ikx}$$

The solutions represent left and right moving waves.

But we have a big problem. The solutions are imaginary exponentials, which means they do not decay. Their wave length is same forever. This is not good.

The probability wave can extend across the whole universe. It means the particle has same probability across the whole universe. This does not make physical sense. Physicists tame mathematical functions by using common sense. If the map on the smartphone is asking you to go off a cliff, you will hopefully use common sense too!

What to do?

The answer lies in adding lots of waves. Here is an example of a plane wave where all wave fronts are similar to one another.

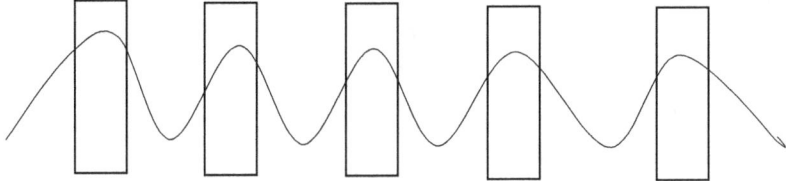

The study of superimposition of waves is well known and is called Fourier analysis.

There is a formula to add waves to get complex wave phenomenon.

$$f = \frac{a_0}{2} + \sum_{n=1}^{\infty} (a_n \cos \frac{n\pi x}{L} + b_n \sin \frac{n\pi x}{L})$$

The Fourier series involves adding various sine and cosine waves. a_0 is like the average height of the wave. We won't go into the details of Fourier series but just remember there is a rigorous way to add waves.

Addition of waves lead to the formation of wave packet.

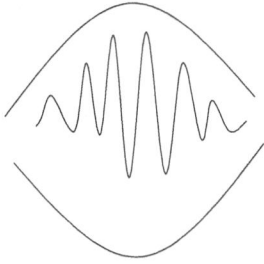

Now we are in business. There is localization of the wavefunction. The wavefunction can be normalized. But it comes at a cost. Each wave has different frequency. There are lot of frequencies in the wave packet. Frequency means how fast a wave is wiggling. The frequency is related to the wave number k.

$k = 2\pi f$ or $\frac{2\pi}{\lambda}$

The wave number represents how many waves can be packed over 2π. A wave completes its full circle over 2π.

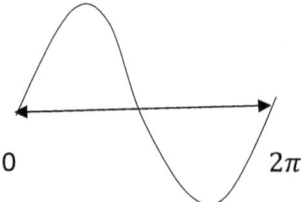

The wave number is 1. But if we can fit three waves into it then wave number will be 3.

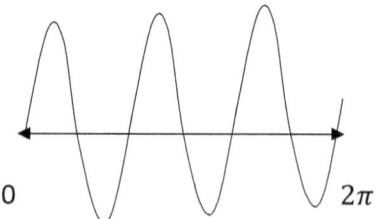

The wavenumber is closely related to momentum.

$p = \hbar k$

The logic is simple. If you make a fist, you are packing fingers of different length. The punch is much stronger than a slap!

The wave packet has waves of different momenta.

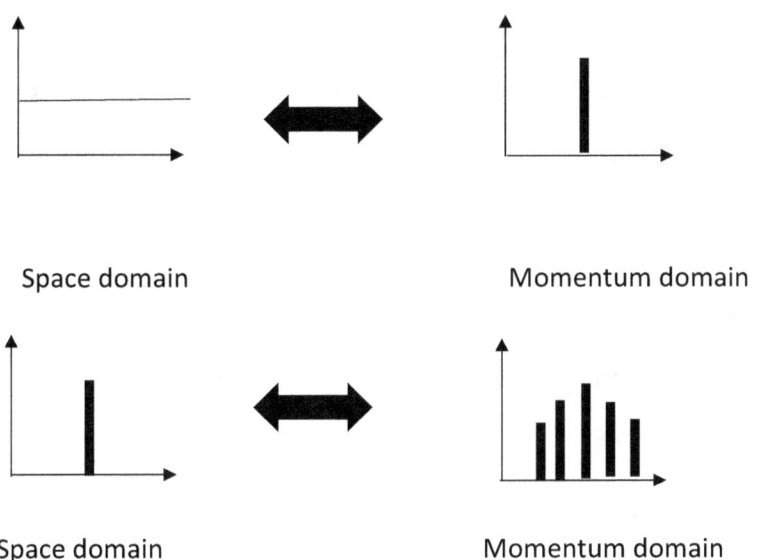

Space domain   Momentum domain

Space domain   Momentum domain

As you can see, space and momentum domains behave oppositely. When a particle is localized, momentum is uncertain and vice versa. This is the basis for the famous Heisenberg's uncertainty principle.

The two domains represent the same information, only the emphasis is different. Think of x ray and ultrasound. X ray shows bones clearly and the ultrasound is good for soft tissues. Both are looking at the same body organ.

This allows for the space domain to be converted into momentum domain and vice versa. The procedure is called Fourier transformation.

$\psi_x \Leftrightarrow \tilde{\psi}_k$

$\psi_x = \frac{1}{\sqrt{2\pi}} \int \tilde{\psi}_k e^{ikx} dk \Leftrightarrow \tilde{\psi}_k = \frac{1}{\sqrt{2\pi}} \int \psi_x e^{-ikx} dx$

The subscript over the momentum wave function is customary and is called a twiddle.

Velocity of the waves

There are two kind of velocities. The phase velocity is the speed of the individual waves inside the wave packet. The group velocity is the velocity of the wave packet.

The speed of a wave $= \frac{distance}{time} = \frac{\lambda}{T}$

The wave number $k = \frac{2\pi}{\lambda}$

The angular frequency $\omega = \frac{2\pi}{T}$

It tells how fast the wave is completing its cycle.

In terms of w and k,

$v = \frac{\omega}{k}$

This is called the phase velocity.

In case of light, we know the speed of light is always constant.

$c = \frac{\omega}{k}$

Each light wave moves with the speed c. This also means that the group velocity is also c as there is no difference in the speed of individual waves. Everything in light moves with speed c.

In case of light or electro-magnetic waves, group velocity-=phase velocity=$\frac{\omega}{k}$

$\frac{\omega}{k}$ is also called the dispersion relation. It tells us how a wave packet gets distorted. For non-light waves, the phase velocity is not constant. Each wave is moving with different velocity. This can cause the wave packet to change shape and get distorted.

The phase velocity is easy to get from the wave solution.

$$y = A\sin(kx - wt)$$

Let me spend some time to tell you how we get the wave equation.

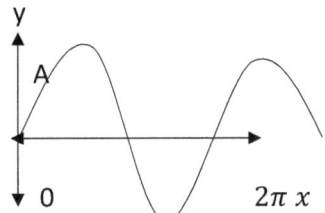

As you can see the amplitude A on the y axis repeats itself after $\frac{2\pi}{\lambda}$ distance on the x axis. This is like θ. The function can be sin or cosine, doesn't matter as the difference is only in the starting point.

$$y = A\frac{2\pi}{\lambda}x = A\sin\theta \text{ or } A\sin kx$$

What if the wave is moving to the right?

Let's take a real-world example. You measure the length of a playing field to be 100m. A person who is going at 10 m/s speed to one end of the field will measure the field to be 100-10=90 m at one second mark. For him the length of the field is $x - vt$.

Similarly, for a wave travelling to the right, the solution changes by

$$y = A\sin(kx - wt)$$

The sine wave can be written in the exponential form using Euler's formula

$$y = Ae^{i(kx-\omega t)}$$

For plane waves $kx - \omega t$ is constant as each wave fronts are similar to one another, moving like compartments of a train.

$$C = kx - \omega t$$

$$kx = C + \omega t$$

$$x = C + \frac{\omega}{k}t$$

Doing differentiation leads to

$$\frac{dx}{dt} = 0 + \frac{\omega}{k}$$

$v = \frac{\omega}{k}$, this is the phase velocity.

Going back to the solution of a free particle

$$\psi = c_1 e^{i(kx-\frac{E}{\hbar}t)} + c_2 e^{-i(kx+\frac{E}{\hbar}t)}$$

$$v = \frac{\omega}{k} = \frac{E}{k\hbar}$$

Using $k = \frac{\sqrt{2mE}}{\hbar}$ or $E = \frac{k^2\hbar^2}{2m}$, we get $w = \frac{\hbar k^2}{2m}$ and

$$v = \frac{\hbar k}{2m}$$

This is the dispersion relation for matter waves.

What about the group velocity?

We will assume that the packet is compact, which means it has a dominant central frequency ω_0 and wave number k_0

The individual frequencies are only a little bit away from the central value.

In order for waves of different k to form a bump, their phases have to add constructively. We can say the phase $kx - \omega t + \varepsilon$ between various waves does not depend on k. Even when we form a fist, different fingers come together as they are bent in the same way. We cannot form a fist if each finger is doing its own thing.

ε is additional phase, represents the difference in starting point of waves.

$$\frac{d\,phase}{dk} = 0 \text{ or } \frac{d\,(kx-\omega t+\varepsilon)}{dk} = x - \frac{d\omega}{dk}t = 0$$

$$\frac{x}{t} = \frac{d\omega}{dk} = \text{group velocity.}$$

The group velocity is $\frac{d\omega}{dk}$

The group velocity expression makes sense as change from the central value measures the group velocity. If there is no change and each wave have the same w/k then group velocity is same as phase velocity as seen for EM waves.

$$v_g = \frac{d\omega}{dk} = \frac{d(ck)}{dk} = c \text{ and } v_p = \frac{w}{k} = c \text{ for light waves.}$$

$$v_g = \frac{d\omega}{dk} = \frac{d\left(\frac{\hbar k^2}{2m}\right)}{dk} = \frac{\hbar k}{m} \text{ and } v_p = \frac{\hbar k}{2m} \text{ for matter waves}$$

The classical velocity $E = \frac{1}{2}mv^2$ or $v = \sqrt{\frac{2E}{m}}$

Substituting $E = \frac{\hbar^2 k^2}{2m}$ as we first defined k as $\frac{\sqrt{2mE}}{\hbar}$, will give

$$v = \frac{\hbar k}{m}$$

So classical velocity=group velocity.

The Fourier transform can also be used to derive the group velocity for the generalized case, but I think we should end here. I hope you have got the point.

Think of a glass of cold beer, the water molecules are moving rapidly inside the mug. This is like the phase velocity. As you pick the mug to get a sip of the beer, the movement of the mug is like the group velocity.

Uncertainty Principle

Heisenberg's uncertainty principle is the flag bearer of quantum mechanics. The uncertainty principle along with Schrodinger's cat has seeped into the popular culture. It is a different matter that most people have no idea what it really means. So, let's explore it in detail as it is key to understanding quantum mechanics.

I have you have a good understanding of randomness now. All quantum results are random. You can say that this makes results uncertain, but this is not what uncertainty principle refers to. Going over the randomness again, if you have a bunch of quantum particles in the same state e.g. in the ground state and you want to know the location of the electron around the nucleus then every time you check the answer, it is random. The answer from 1 to say 100 atoms will be random. But once you get the result, it is certain. Let's say you do the experiment on atom number 1 and found the electron at distance d from the nucleus. If you repeat this experiment on atom number 1 millions of times, you will get the same answer. Once the measurement is done, superposition is gone, and random results are in, there are no more surprises, the results are final. The measurement on atom 2 may have a different answer which is also random but once the measurement is done, it is certain and does not change.

What is uncertainty then?

The uncertainty refers to the uncertain results that we get when we try to measure different things on the same quantum state. Once we find the exact position of the ground state electron on the atom 1, then we try to measure the momentum of the electron, the results are completely uncertain. The electron is moving in all directions. It does not seem right, you think we need to design a different experiment on the atom to know the momentum. You are successful in knowing the momentum of the electron. Then you go back again and check the position of the electron and to your surprise, the electron is not there. This is the uncertainty principle. It refers to the measurement of one quantity affecting the measurement of another quantity. The result is that we do not have full information about the quantum state. Either we know the position or the momentum, not both. There is always uncertainty about the full picture of the quantum state. The uncertainty is not just confined to position and momentum, there are other quantities that uncertainty principle is applied e.g. only one component of angular momentum and spin of the electron can beknown. If you try to measure the other component of the spin, the first one becomes uncertain.

How do we describe it mathematically?

First measure position then measure momentum, in short xp

Or first measure momentum then measure position px

If measuring position has no effect on measuring momentum and vice versa,

then $xp - px = 0$

This is called commutation relation and is mathematically described as

$[x, p] = xp - px$

$[p, x] = px - xp$

Since position and momentum measurement affect each other then

$[x, p] = xp - px \neq 0$

I don't know if you are getting what's going on, this is downright weird. If I look at a person walking, I exactly know where that person is and where is he going. But this assumes that process of looking itself is a bystander process and does not affect the outcome. The photons that strike the walking person and reach my eye do not affect the outcome. But this does not happen in quantum mechanics. It is like we are looking for a person in the dark. The only way to find out the position is by throwing a football forcefully where a person could be found and if it bounces back then we got our position. But once the football hits the person, he falls off in some direction making the momentum uncertain. This is what happens in quantum mechanics.

 Catch the football or the person, choice is yours.

The commutation relation should not be too surprising as we are dealing with operators, not numbers.

$7 \times 8 - 8 \times 7$ is obviously zero but x and p are not numbers.

Think of this way, studying and exam results can be considered operators, so

Study × Exam − Exam × Study is not zero. If you give exam first and then study, there will be uncertainty in your career prospects!

If the measurement results are uncertain as they are being affect by measurement of other quantities on the single quantum state, you will get lot of numbers. We need statistics to make sense of the uncertain results.

The obvious statistical method is to calculate the mean. If we have 5,5,10,2,3 as the measurement results, the mean is

$$\frac{5+5+10+2+3}{5} = 5$$

The mean or the average gives an idea where the value of most numbers lies.

But it's not a perfect, these numbers have the same mean as well

$$\frac{22+(-5)+5+1+2}{5} = 5$$

A single large number can skew the mean. We need a better measure. The variance is a better way as it tells us the spread of the results.

The simplest way to calculate the spread will be to subtract mean from each value. This is the range of numbers.

$n - mean$ is the range but if you add them, it doesn't work as higher and lower values cancel each other e.g. in case of first example

(5-5) + (5-5) + (10-5) + (2-5) + (3-5) =0

To avoid this problem, we square the terms and divide, this is called variance.

$$\sigma^2 = \frac{(n-mean)^2}{n}$$

In the above case, variance will be

$$0 + 0 + 25 + 9 + 4 = \frac{38}{5} = 7.6$$

The variance in this case is more than the mean, it suggests that spread of values is more than what the mean would suggest which makes sense as the lowest value 2 and highest value 10 are quite apart than the mean.

Taking the square root gives the standard deviation

$$\sigma = \sqrt{\sigma^2} = 2.8$$

It gives you an idea how far apart are most values from the mean.

If you have given exams, you are aware that the exam marks are given in terms of standard deviation from the mean. One standard deviation captures 68.2% of the values in a normal distribution. The large sample of independent processes like students giving exams usually form a normal distribution or a Bell curve where most values are spread equally around the mean. A standard normal distribution is formed by mean of 0 and standard deviation of one.

Quantum mechanics is about the operators or observables, so let's put them in the variance.

First, the mean is the expectation value and is denoted by $\langle n \rangle$

The variance of position will be

$$\sigma_x^2 = (x - \langle x \rangle)^2$$

$$\sigma_x^2 = (x - \langle x \rangle)(x - \langle x \rangle)$$

Let's call $(x - \langle x \rangle) = A$

If we let it act on the wave function and take the inner product, we have

$\langle A\Psi | \Psi A \rangle$

Due to normalization $\langle \Psi | \Psi \rangle = 1$, so

$$\sigma_x^2 = \langle A | A \rangle$$

Similarly, for momentum

$$\sigma_p^2 = \langle B | B \rangle$$

What is A?

It is a complex number. The complex numbers are vectors e.g.

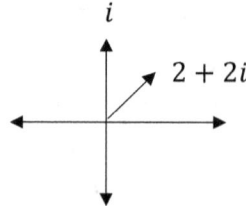

We can form a triangle with vectors.

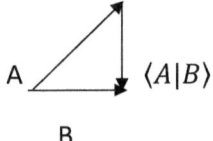

B

$\langle A|B \rangle$ is the projection of the vectors and thus forms the part of the triangle. We will use the logic that two sides of a triangle are always greater than or equal to the third side. This is expressed in the Cauchy- Schwarz inequality.

$\sigma_x^2 \sigma_p^2 = \langle A|A \rangle \langle B|B \rangle \geq \langle A|B \rangle^2$

Or $\sigma_x^2 \sigma_p^2 \geq \langle A|B \rangle^2$

There is another theorem to use

The complex number magnitude is given by

$|z|^2 = zz^* = (x+iy)(x-iy) = x^2 + y^2$

$|z|^2 = (real)^2 + (imaginary)^2 \geq (imaginary)^2$

This theorem just says whole of something is more than part of something!

The real part x is $\frac{z+z^*}{2}$ as $\frac{x+iy+x-iy}{2} = x$

Likewise, imaginary part $y = \frac{z-z^*}{2i}$ as $\frac{x+iy-x+iy}{2i} = y$

$|z|^2 \geq \left(\frac{z-z^*}{2i}\right)^2$

Let's put the operators for the complex numbers.

Let $\langle A|B\rangle = z$

$$|\langle A|B\rangle|^2 \geq \left(\frac{\langle A|B\rangle - \langle B|A\rangle}{2i}\right)^2$$

Since $\sigma_x^2 \sigma_p^2 \geq \langle A|B\rangle^2$, we have

$$\sigma_x^2 \sigma_p^2 \geq \left(\frac{\langle A|B\rangle - \langle B|A\rangle}{2i}\right)^2$$

We need to calculate $(\langle A|B\rangle - \langle B|A\rangle)^2$

Putting back the value of A and B and acting on wave function, we will get

$\langle A|B\rangle = ((x - \langle x\rangle)\Psi | p - \langle p\rangle \Psi))$

There will be term like $\langle \Psi | xp\Psi\rangle = \langle xp\rangle$

The terms like $\langle \Psi | x\langle p\rangle \Psi\rangle$ will simplify to $\langle p\rangle \langle \Psi | x\Psi\rangle = \langle p\rangle \langle x\rangle$

Note by definition $\langle \Psi | x | \Psi\rangle = \langle x\rangle$

The expectation value $\langle p\rangle$ is a number and we can move it around but be careful with order of the operators which cannot be changed.

There will be some cancellation of terms and we will get

$\langle A|B\rangle = \langle xp\rangle - \langle x\rangle \langle p\rangle$

Similarly,

$\langle B|A\rangle = \langle px\rangle - \langle x\rangle \langle p\rangle$

$\langle A|B\rangle - \langle B|A\rangle = \langle xp\rangle - \langle px\rangle = \langle [x,p]\rangle$

So, in the end we get

$$\sigma_x^2 \sigma_p^2 \geq \left(\frac{1}{2i}\langle [x,p]\rangle\right)^2$$

This is the generalized form of uncertainty principle. It is not confined to position or momentum. Any pair of operators can be used.

For the position, momentum case, we need to calculate their commutator.

$$[xp] = x\frac{\hbar}{i}\frac{\partial \Psi}{\partial x} - \frac{\hbar}{i}\frac{\partial}{\partial x}(x\Psi) = \frac{\hbar}{i}(x\frac{\partial \Psi}{\partial x} - x\frac{\partial \Psi}{\partial x} - \Psi\frac{\partial x}{\partial x}) = -\frac{\hbar}{i}\Psi = i\hbar\Psi$$

Putting in the value of the commuter in the uncertainty principle

$$\sigma_x^2 \sigma_p^2 \geq \left(\frac{1}{2i}i\hbar\right)^2 = \left(\frac{\hbar}{2}\right)^2$$

$$\sigma_x \sigma_p \geq \frac{\hbar}{2}$$

This is *Heisenberg's uncertainty principle*.

It states that when position and momentum are measured, there is always uncertainty involved and there is limit to the maximum precision that can be achieved, given by $\frac{\hbar}{2}$.

Is uncertainty principle simply about measurement?

You may get the impression that if measurement can be gentle and does not affect the outcome, we can achieve maximum precision where no variance is produced. So, it is just a matter of building precise instruments. But it is an erroneous argument. Remember, when we solved the Schrodinger equation, the solution was only possible when we produced a wave packet by Fourier analysis. The mathematical basis of position and momentum requires uncertainty. There is no way around it.

A longer wavelength will miss the position of the particle but will not give it a kick. A high frequency wave with high energy will locate the position of the particle precisely but will kick the particle hard, making the momentum uncertain.

How does expectation value changes with time?

$\frac{d\langle m \rangle}{dt}$, let's have the full definition of the expectation value and then do differentiation.

$$\frac{d\langle \Psi | m | \Psi \rangle}{dt}$$

Doing product differentiation will give three terms

$$\left\langle \frac{\partial \Psi}{\partial t} \middle| M\Psi \right\rangle + \left\langle \Psi \middle| \frac{\partial M}{\partial t} \Psi \right\rangle + \left\langle \Psi \middle| M \frac{\partial \Psi}{\partial t} \right\rangle$$

Replace $\frac{\partial \Psi}{\partial t}$ using the Schrodinger equation

$$\left| \frac{\partial \Psi}{\partial t} \right\rangle = \frac{1}{i\hbar} H\Psi$$

$$\left\langle \frac{\partial \Psi}{\partial t} \right| = -\frac{1}{i\hbar} H\Psi$$

$$-\frac{1}{i\hbar} \langle H\Psi | M\Psi \rangle + \left\langle \frac{\partial M}{\partial t} \right\rangle + \frac{1}{i\hbar} \langle \Psi | MH\Psi \rangle$$

Using the property of Hermitian operators, H can be moved past the wavefunction in the first term

Also $\frac{1}{i\hbar} \times \frac{i}{i} = -\frac{i}{\hbar}$

So, we will have

$$\frac{i}{\hbar} \langle \Psi [HM - MH] \Psi \rangle + \left\langle \frac{\partial M}{\partial t} \right\rangle$$

$$\frac{d\langle m \rangle}{dt} = \frac{i}{h}\langle [H,M] \rangle + \left\langle \frac{\partial M}{\partial t} \right\rangle$$

The expectation value does not change if the operator commutes with the Hamiltonian and does not explicitly depends on time.

That's just too technical. Let me use simpler language. Suppose a ball is sitting on the table. $\frac{d\langle m \rangle}{dt}$ means we want to know the average position of the ball at a later time e.g. after 1 hr. Explicit time dependence $\left\langle \frac{\partial M}{\partial t} \right\rangle$ of the operator M means if for example we hit the ball every hour, then it is explicitly time dependent. If there is no kick to the ball and we just want to know the position of the ball at a later time, then $\left\langle \frac{\partial M}{\partial t} \right\rangle$ term will disappear.

If Hamiltonian and the operator M commutes meaning the measurement of one does not affect the measurement of the other, then the expectation value does not change with time. In our example of a ball sitting on the table, it has some potential energy based on the height of the table, kinetic energy is zero as it is still. In this case, you can measure the height of the ball (potential energy) or the position of the ball, it does not matter. This is to say position and Hamiltonian commute. So, the ball is just sitting there. You can check it 1 hour later, it will be found at the same position. Technically you say the expectation value is not changing with time.

However, if we kick the ball every hour, then it's obvious that ball will not be found at the same place at a later time, so its expectation value will not be conserved. Measuring position then kicking the ball is not the same as kicking the ball and measuring position. This means position and Hamiltonian do not commute. I hope you are getting the idea behind the mathematics.

Let's use the time dependence of the expectation value of the operator to modify the uncertainty principle.

$$\sigma_H^2 \sigma_M^2 \geq \left(\frac{1}{2i} \langle [H,M] \rangle \right)^2 \text{ and} \langle [H,M] \rangle = \frac{h}{i} \frac{\partial \langle M \rangle}{\partial t}$$

$$\sigma_H^2 \sigma_M^2 \geq \left(\frac{1}{2i} \frac{h}{i} \frac{\partial \langle M \rangle}{\partial t} \right)^2 \text{ or}$$

$$\sigma_H \sigma_M \geq \frac{\hbar}{2} \frac{\partial \langle M \rangle}{\partial t}$$

If we let $\Delta E = \sigma_H$ and $\Delta t = \frac{\sigma_M}{\frac{\partial \langle M \rangle}{\partial t}}$

Then $\Delta E \Delta t \geq \frac{\hbar}{2}$

This is called the *energy time uncertainty principle*. It is very different than the Heisenberg's uncertainty principle. ΔE refers to spread in the energy values or uncertainty in the energy measurements, that's straight forward. But the meaning of Δt is less clear. It refers to a significant change in the mean value of the measurements. In our example of the ball on the table, if the ball falls down to the ground, the expectation value of the position of the ball would change significantly. The time that it takes for the ball to fall to the ground would be Δt.

If the expectation value is changing fast, Δt is small and vice versa. If you want accurate energy measurements or ΔE to be small, Δt has to be large or expectation value should be changing slowly. The stationary states that do not change with time ($\frac{\partial \langle x \rangle}{\partial t} = 0$), we can do accurate energy measurements.

On the other hand, if expectation value is changing fast, ΔE would be large. This means we are uncertain about the energy of different states. This can lead to interesting consequences. The most notable being an apparent violation of the classical law of energy conservation. This violation can happen over a very short period of time which is not experimentally detectable. If we are not sure about the energy of the ball on the table and on the floor, then the ball on the floor can make a quick trip to the table and back without any intervention. This would violate the energy conservation as ball on the floor has lower potential energy than the ball on the table.

A virtual round trip!

We can say the ball made the virtual trip as long as Δt is very small. But this seems like a ridiculous assertion as you cannot detect it experimentally. But this is what happens in particle physics. The virtual particles are created out of vacuum, violating energy conservation. But since they last for very short period of time, they are not detectable. They do have indirect affects when calculating the probabilities of various particle physics experiments. We will study the virtual particles in detail in later chapters.

Link to Classical Physics

We can't just abandon classical physics. It works brilliantly in our daily like. Most of the rocket science is classical physics. But quantum mechanics is the fundamental theory. It should lead to classical physics formulae. At first, it seems like a hopeless exercise as quantum mechanics is so different than classical physics. It is a reasonable assumption to make that if you look very closely at a thing, you get quantum mechanics but if you look from a distance, classical physics should be seen. We should expect that classical physics is quantum mechanics on average.

The classical physics is based on Newton's law of motion or F=ma.

The work done, or energy used $U = -\int F dx$

$F = -\frac{dU}{dx}$

We know from classical physics that rate of change of momentum is force. The expectation value of momentum is good candidate to get the classical formula.

$$\frac{d\langle p\rangle}{dt} = \frac{i}{\hbar}\langle[H,p]\rangle$$

The total energy or Hamiltonian $= \frac{p^2}{2n} + U$

If we put its value in the commutator then

$[p^2, p] = 0$, as momentum obviously commutes with itself. We are left with U only.

$[U, p] = Up - pU = ?$

Let's put the operator of momentum and act it on the wavefunction.

$$U\frac{\hbar}{i}\frac{\partial \Psi}{\partial x} - \frac{\hbar}{i}\frac{\partial(U\Psi)}{\partial x}$$

Doing product differentiation

$$U\frac{\hbar}{i}\frac{\partial \Psi}{\partial x} - \left(\frac{\hbar}{i}\Psi\frac{\partial U}{\partial x} + U\frac{\hbar}{i}\frac{\partial \Psi}{\partial x}\right)$$

We are left with $-\frac{\hbar}{i}\Psi\frac{\partial U}{\partial x}$

But we need to put the above value in $\frac{i}{\hbar}\langle\Psi[H,p]\Psi\rangle$ expression. We already used one wavefunction above, the other one has to be put as well, so

$$\frac{i}{\hbar}\left(-\frac{\hbar}{i}\Psi\frac{dU}{dx}\right)\Psi \text{ or } \left\langle-\frac{dU}{dx}\right\rangle$$

This resembles the expression of the classical physics. It is called the Ehrenfest theorem.

What about $\frac{d\langle x\rangle}{dt}$?

Assuming no potential and only kinetic energy,

$$\frac{d\langle x\rangle}{dt} = \frac{i}{\hbar}[H,x] = \frac{i}{\hbar}\left[\frac{p^2}{2m},x\right]$$

Using commutator algebraic rules

$$[p^2,x] = p[p,x] + [p,x]p$$

We already know that $[p,x] = i\hbar$

$$\frac{i}{2m\hbar}[-2i\hbar p] = \left\langle \Psi \left|\frac{p}{m}\right| \Psi \right\rangle = \left\langle \frac{p}{m} \right\rangle$$

In classical physics, p=mv or $v = \frac{p}{m}$

Again, the average classical velocity resembles the time derivative of the expectation value of position.

This is not the end of similarities between classical physics and quantum mechanics. The commutation relations in quantum mechanics have their classical counterpart as well. They are called Poisson brackets.

$$\{A,B\} = \frac{\partial A}{\partial x}\frac{\partial B}{\partial p} - \frac{\partial B}{\partial x}\frac{\partial A}{\partial p} \text{ e.g.}$$

$$\{x,p\} = \frac{\partial x}{\partial x}\frac{\partial p}{\partial p} - \frac{\partial p}{\partial x}\frac{\partial x}{\partial p} = 1-0 = 1$$

We know that $[x,p] = i\hbar$

The relationship between commutator and Poisson bracket is

$$[x,p] = i\hbar\{x,p\}$$

The commuters and Poisson brackets have similar mathematical relations

$$[x,p] = -[p,x] \qquad \{x,p\} = -\{p,x\}$$

$$[x+y,p] = [x,p]+[y,p] \qquad \{x+y,p\} = \{x,p\}+\{y,p\}$$

I do not want to do over all the properties but they part of the so-called Lie algebra.

If $\{f, H\}=0$ then f is a conserved quantity similar to the quantum mechanics.

The time change of the function also mimics the time change of the expectation value in quantum mechanics.

$$\frac{d}{dt}f = \{f, H\} + \frac{df}{dt}$$

I think this is enough PB for you. The upshot is that mathematical structure of Hamiltonian classical physics is similar to quantum mechanics even though the mathematics take on a deeper meaning in quantum mechanics.

Path Integral Method

It is an alternative way to explain quantum mechanics. The path integral equation is equivalent to the Schrodinger equation. The method was developed by Paul Dirac and Richard Feynman. The Schrodinger equation is based on the Hamiltonian method. The total energy (K.E. +P.E.) determines how particle moves from one point to another along a path. The path integral method is based on Lagrangian. K.E.- P.E. forms the Lagrangian. The most efficient path that minimizes the Lagrangian is selected as the correct path. This is called the action.

Action or $S = \int L dt$

Propagator or $U = C \sum e^{\frac{i}{\hbar}S}$

The propagator is made of the sum of all paths. The propagator takes the wave function from time 0 to time t.

$\Psi_0 U \rightarrow \Psi_t$

Each path has the same probability amplitude but different phase. It is compatible with the uncertainty principle as we are not sure of the exact path.

This looks like a hopeless exercise. If particle has to go from A to B, we have to sum over all paths that have equal probability amplitude. It looks like an impossible task.

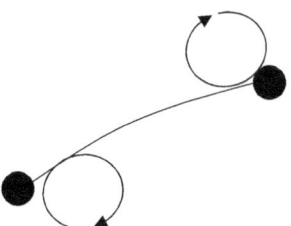

The phases of the paths come to the rescue. Most of the convoluted path phases cancel each other and we are left with the path which is the closest distance between A and B and thus the most probable result.

It seems that nature is thrifty and efficient. It only selects the path that minimizes the action or difference between K.E. and P.E.

The Hamiltonian and Lagrangian methods are entirely equivalent in classical physics. Similarly, the Schrodinger equation and path integral methods are equivalent as well. If you put in the value of the Lagrangian and carry out the integration, Schrodinger equation comes out in the end.

There is a philosophical question here.

Does the particle already know its destination and then selects the appropriate path from available choices?

The sum is over the histories. It is a mathematical method to get the most probable answer. The particle that originates from the sun does not know at the start that it will reach your car window to cause the glare. There can be other explanations, but

I would stick with the most conservative one that we are doing the calculation backwards from the things that already happened.

What is it good for?

The propagators and Lagrangians are very important to describe quantum field theory, which is a deeper theory than elementary quantum mechanics. It is compatible with Einstein's special theory of relativity. We will study it in later chapters.

Chapter 4

Simple Quantum Models

Particle in a Box

It is an ideal situation, physically unrealistic but it is a good model to show quantum effects.

The particle is boxed in.

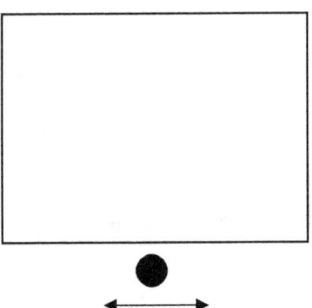

There are certain assumptions to be made for this model

1. The length of the box is L
2. The potential energy at the walls is infinity. It is like saying the walls are infinitely high.
3. The potential energy inside the box is zero.

4. The particle is stuck inside the box. The wave function has to end at the walls. It ends smoothly at the walls.
5. The particle only has kinetic energy inside the walls.

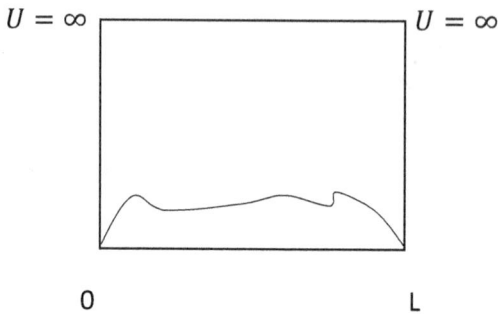

With this information, we can solve the Schrodinger equation.

$H\Psi = E\Psi$

The Hamiltonian inside the box

$H = K.E. = \frac{1}{2}mv^2 = \frac{p^2}{2m}$ (p=mv)

Since Hamiltonian has no time dependence, we use the simpler time independent Schrodinger equation. After all, our goal is to find the measurement values or eigenvalues of the Hamiltonian.

To find the eigenvalues, we need to convert the equation in terms of momentum operator $-i\hbar \frac{d}{dx}$

$\frac{-\hbar^2}{2m} \frac{d\Psi^2}{dx^2} = E\Psi$

$$\frac{d\Psi^2}{dx^2} = -\frac{2mE}{\hbar^2}\Psi$$

Let's simplify the equation by introducing new variable $k = \frac{\sqrt{2mE}}{\hbar}$

$$\frac{d\Psi^2}{dx^2} = -k^2\Psi$$

The solution to the above equation is straight forward. Whenever you get back the same function Ψ even after differentiation, you think of exponentials.

$$\frac{de^u}{dx} = \frac{du}{dx}e^u$$

$$\Psi = c_1 e^{ikx} + c_2 e^{-ikx}$$

The solution is made of right and left moving waves.

Equivalently, using Euler's formula $e^{ikx} = coskx + i\,sinkx$ and $e^{-ikx} = coskx - i\,sinkx$

The solution can be rewritten as

$\Psi = A\,sin\,kx + B\,coskx$

We will use this version here.

We want to use boundary conditions to know the value of constants.

At x=0, $\Psi = 0$

As $cos\,0 = 1$, cos is out and B=0.

$sin0 = 0$. This is ok.

At the other end, x=L, $\Psi = 0$

$AsinkL = 0$

Either A=0 or SinkL =0. If we make A=0 then $\Psi = 0$ everywhere which is no good.

We let $sinkL = 0$

Sine function is zero only if it 0 or multiple of π

$sinn\pi = 0$

$kL = n\pi$

$k = \frac{n\pi}{L}$

Now $k = \frac{\sqrt{2mE}}{\hbar}$

We get $E = \frac{n^2\pi^2\hbar^2}{2mL^2}$

This is an important result. The energy comes in integer multiples. The energy cannot take any value but only certain values. The energy is quantized.

It should not be too surprising that energy values are restricted. When you want to fit only certain number of waves in a box, you either fit the whole wave or none at all. This quantizes energy. The more waves that can be fit in, the higher the frequency and the energy as $E = \hbar f$. This is what we do in gym too, right? The faster we jump, the more energy we burn!

$\Psi_n = A \sin \frac{n\pi}{L}$

To find A, use the normalization condition

$\int |\Psi|^2 dx = 1$, the integration is all over space, particle has to be found somewhere.

$\int_0^L \left(A \sin \frac{n\pi x}{L}\right)^2 dx = 1$

The solution to this integral is $\frac{L}{2}$.

$$A^2 \frac{L}{2} = 1 \text{ or } A = \sqrt{\frac{2}{L}}$$

$$\Psi_n = \sqrt{\frac{2}{L}} \sin \frac{n\pi}{L}$$

What to do with Ψ_n?

It tells us the probability of finding the particle in a box for various n's or energies.

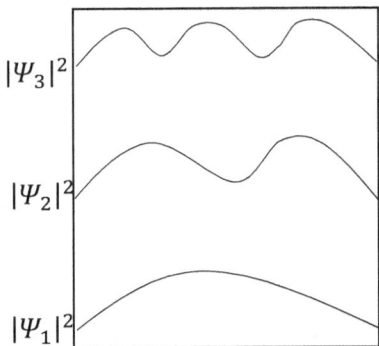

What can we conclude?

1. Ψ_1 is the ground state. It is the lowest energy. The energy cannot be zero in keeping with uncertainty principle. If energy was zero, particle will be sitting, and we can know position and momentum precisely.

2. In the ground state, the particle is most likely to be found at the center of the box. This is strange, if it was a soccer ball going back and forth in a box with constant speed, it should be found equally everywhere along the box.

3. When we look at an excited state like Ψ_2, there is a place in the box where particle is never going to be found. This is again weird. The zero-probability spot is called the node. This all makes sense if we think of particle as a wave and nodes are areas of destructive interference and antinodes are areas of positive interference. But we cannot get away from quantum weirdness. What is interfering with what? There is only one particle and somehow it interferes with itself.

4. As we keep increasing energy or n, the probability density flattens out and it resembles the classical probability.

5. There is no time dependence to the probability states. The probability of find the particle is same after 5 seconds or 5 years. These are called stationary states. They are the solutions of the time independent Schrodinger equation and time independent potential.

6. Quantum mechanics is a probabilistic theory. It means that we have to do lots of experiments and collect data in order for the theory and the experiments to match each other. If we do a single experiment, we will find the particle somewhere in the box. We can naively assume that particle was just there. If we keep checking the same box again and again, we will find the particle at the same place. This is because superposition has collapsed, and a random experimental result has come out. To see that you have to do repeated experiments on identical boxes. We call them same undisturbed states. Once we measure particle in a box, that state is disturbed. We have to move to another box to see the probability picture. Since all the boxes are identical, we should find the particle in same spot in all the boxes. This does not happen. If we plot a graph where we find particle in each box, the probability distribution will match exactly with the theoretical results. You may be thinking how can we prepare the boxes to be in the same state? Well, it is a contrived example. In reality, you do experiment on a specific quantum state e.g. ground state of hydrogen and then measure the probability of the

electron. You will find the graph that matches the solution of the Schrodinger equation.

Similar situation happens in classical probability too. We know that the probability of getting heads or tails in a coin is 50:50. But if you throw 10 coins, it's possible you get 7 heads and 3 tails. It is only after doing millions of experiments that the probabilities will approach 50:50. Probability theory requires lot of hard work, keep on repeating the experiments.

But keep in mind that in classical probability, it is theoretically possible to get the result you want. May be a trained coin flipper knows how to throw a coin to get heads or tails, so the process is not random. But quantum probability is truly random. When we measure where the particle will be in its ground state, the only thing certain is that it will be somewhere in the plotted graph. But which spot in the graph will come out when we do an experiment cannot be predicted, its random.

$$\Psi = c_0 \Psi_0 + c_1 \Psi_1 + c_2 \Psi_2 = \Sigma c_n \Psi_n$$

The solution of the particle in a box is made up of ground state Ψ_0, first excited state Ψ_1 and so on. Each state has certain probability of occurring and is given by $|c|^2$. Only on repeating the experiments and collecting data, we will find the probability of getting ground state or excited state etc. Once we have found a state then there is corresponding probability graph where the particle can be found.

It is similar to flipping a coin.

$$\Psi_{coin} = c_H \Psi_H + c_T \Psi_T$$

$$c_H = c_T = \frac{1}{\sqrt{2}}$$

$$|c_H|^2 = |c_T|^2 = \frac{1}{2}$$

Finite Potential

Let's make particle in a box more realistic, by making potential energy very high but finite.

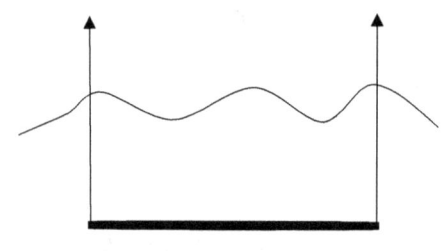

L R

The solution inside the box is unchanged

$$\Psi_{in} = A\sin kx + B\cos kx$$

The boundary equations are not the same. The wave function is not zero at the boundary, so we cannot solve the constants. We need to compare the solutions outside and inside the box.

The Schrodinger equation outside the box is

$$\frac{d^2\psi}{dx^2} = \frac{2m(U-E)}{\hbar^2}\psi$$

We have assumed that potential U is higher than the energy of the particle.

Introducing new variable $\alpha = \frac{\sqrt{2m(U-E)}}{\hbar}$

$$\frac{d^2\psi}{dx^2} = \alpha^2\psi$$

The solution is again exponentials

$\Psi_L = L_1 e^{\alpha x} + L_2 e^{-\alpha x}$ is the solution on the left side of the box.

$\Psi_R = R_1 e^{\alpha x} + R_2 e^{-\alpha x}$ is the solution on the right side of the box.

$L_2 e^{-\alpha x}$ part is rejected. If we go to the left of the box or $-x$ direction, the solution blows up, $L_2 e^{-\alpha \cdot -x} = L_2 e^{\alpha x}$

This type of solution is rejected as it is physically unrealistic. Our particle is in the box, but this solution says that particle would rather be as far away as possible from the box, which is non-sense.

For similar reasons, $R_1 e^{\alpha x}$ is rejected.

To connect the three regions, we make some assumptions.

1. The wave function remains smooth at intersections.

2. The wave function cannot have sharp turns at intersections. This means the rate of change of wave function (first derivative) is equal at intersections.

$\Psi_{L(at\ length\ 0)} = Asink0 + Bcosk0$ and $\Psi_{R(at\ length\ L)} = AsinkL + BcoskL$

$L_1 e^{\alpha 0} = Bcos0$ and $R_2 e^{-\alpha L} = AsinkL + BcoskL$

$L_1 = B$ and $R_2 e^{-\alpha L} = AsinkL + BcoskL$

The first derivatives are

$\alpha L_1 e^{\alpha 0} = kAcos\ k0 - kBsin0$ and $\alpha R_2 e^{-\alpha L} = kAcos\ kL - kBsinkL$

$\alpha L_1 = kA$ and $\alpha R_2 e^{-\alpha L} = kAcos\ kL - kBsinkL$

These equations can be solved by doing some algebra and the answer will look something like $cotkL$ ~ ratio of k and α.

The upshot is that k and α can only have certain values for the equation to be true. Since both variables represent energy, the quantization of energy is ensured.

The solution is not that interesting as we saw the same thing earlier. But the startling result is that the wave function leaks out of the well.

The wave function leaking out means that there is probability of finding the particle outside the box even though it does not have the energy to overcome the wall potential. This is weird. It is like saying that if you cannot climb a wall then do not worry, keep pushing against the wall and you might get lucky!

The region where the wave function leaks out is called the classical forbidden region. The penetrating distance of the wave function is called delta δ.

The penetrating distance is defined as the distance at which the wave function decays by $\frac{1}{e}$.

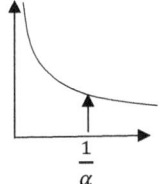

Ψ decays as $e^{-\alpha x}$. We want it to decay to e^{-1}.

as $e^{-\alpha x} = e^{-1}$

or $x = \dfrac{1}{\alpha}$

or $\delta = \dfrac{1}{\alpha} = \dfrac{\hbar}{2m(U-E)}$

This formula makes sense. If particle has higher energy, denominator will be smaller and penetrating distance will increase.

The infinite square well is a special case of the finite potential. If $U = \infty$ then $\alpha = \infty$ and the decaying functions $e^{-\infty}$ will be zero. Only the solution inside the box will remain.

In case you are confused about the decaying and oscillatory functions, here is a quick way.

If E>U, Schrodinger equation is $\dfrac{d^2\Psi}{dx^2} = -\dfrac{2m(E-U)}{\hbar^2}\Psi$ and $\alpha = \dfrac{\sqrt{2m(E-U)}}{\hbar}$

Then $\dfrac{d^2\Psi}{dx^2} = -\alpha^2\Psi$ and the negative sign in front of α makes the solution

$\Psi = A\, e^{i\alpha x} + B\, e^{-i\alpha x}$ which are oscillatory due to the imaginary number i.

If U>E, Schrodinger equation is $\dfrac{d^2\Psi}{dx^2} = \dfrac{2m(U-E)}{\hbar^2}\Psi$ and $\alpha = \dfrac{\sqrt{2m(U-E)}}{\hbar}$

Then $\frac{d^2\Psi}{dx^2} = +\alpha^2\Psi$ and the positive sign in front of α makes the solution

$\Psi = A\,e^{\alpha x} + B\,e^{-\alpha x}$ which are real decaying functions.

Potential Step

Next, we change the situation inside out. The particle is now outside, and it encounters a hurdle. We assume that particle has more energy than the step so that it can easily pass through. If it was a classical particle, we know what would happen. The particle slows down and moves on. It is like a hurdle race where an athlete slows down to jump over the hurdle. Let's see what happens in quantum mechanics.

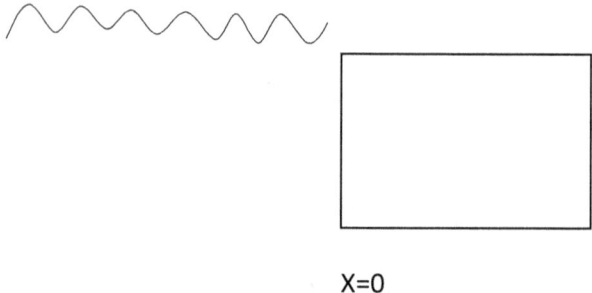

X=0

To the left of the box or x<0, U=0 and the Schrodinger equation is

$$\frac{d\Psi^2}{dx^2} = -\frac{2mE}{\hbar^2}\Psi$$

Variable $k = \frac{\sqrt{2mE}}{\hbar}$

$\frac{d\psi^2}{dx^2} = -k^2\psi$

The solution to the above equation is now familiar to us.

$\psi = Ae^{ikx} + Be^{-ikx}$

The solution is made of right(incoming) and left(reflected) moving waves.

After encountering the box or x>0, the Schrodinger equation is

$\frac{d^2\psi}{dx^2} = -\frac{2m(E-U)}{\hbar^2}\psi$ where $E > U$

Introducing new variable $\alpha = \frac{\sqrt{2m(E-U)}}{\hbar}$

$\frac{d^2\psi}{dx^2} = -\alpha^2\psi$

The solution is again exponentials

$\psi = Ce^{i\alpha x} + De^{-i\alpha x}$

The solution is made of right and left moving waves. The reflection of the wave only happens at x=0 where there is an abrupt change in potential energy. Beyond that, wave is on a higher ground and has no barrier to it. So, there is no reflected or left moving wave beyond x=0. This means the left moving reflected solution $De^{-i\alpha x}$ is rejected.

Probability of incident wave is $|\psi_I|^2$

$\psi\psi^* = A^*e^{-ikx}e^{ikx}A = A^*A = |A|^2$

Note we need complex conjugation for probability calculation. The exponential terms cancel out and since A is a constant, its complex conjugate is also A.

Similarly, probability of reflected wave $|\psi_R|^2 = |B|^2$

And probability of transmitted wave $|\Psi_T|^2 = |C|^2$

Overall reflection probability, $R = \frac{|\Psi_R|^2}{|\Psi_I|^2} = \frac{|B|^2}{|A|^2}$

Overall transmission probability $= 1 - R$

What remains?

We need to find value of the constants. This again involves using the smoothness of the wave function where wave and its first derivative at x=0 are compared.

Ψ at 0, $Ae^{ik0} + Be^{-ik0} = Ce^{i\alpha 0}$ or $A + B = C$ (as $e^0 = 1$)

$\frac{d\Psi}{dx}$ at 0, $ikAe^{ik0} - ikBe^{-ik0} = i\alpha C e^{i\alpha 0}$ or $k(A - B) = \alpha C$

Doing basic algebra will allow you to write constants in terms of variables k and α which in turn can be written in terms of energy and potential.

$R = \frac{(k-\alpha)^2}{(k+\alpha)^2}$

The upshot to all of this is that particle gets reflected unlike a classical particle which either moves ahead or goes back. This just shows that waves properties are different than that of classical particles. This model is useful for scattering which is a study of happens to particles when they collide.

Potential Barrier

If a classical particle encounters a barrier and if it is too high, the particle will hit it and return back. Let's see what a quantum particle does.

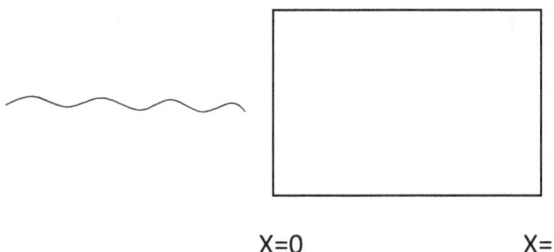

X=0   X=L

I am just going to write the solutions here as we have seen them over and over again.

$X<0, \Psi = A e^{ikx} + B e^{-ikx}$

$X > L, F e^{ikx} + G e^{-ikx}$

As we discussed before, $G e^{-ikx}$ is rejected as there is no potential beyond L to cause any reflection.

Inside the barrier, the Schrodinger equation is

$$\frac{d^2\Psi}{dx^2} = \frac{2m(U-E)}{\hbar^2} \Psi$$

Introducing new variable $\alpha = \frac{\sqrt{2m(U-E)}}{\hbar}$

$$\frac{d^2\Psi}{dx^2} = \alpha^2 \Psi$$

The solution is again exponentials. So, please remember the exponential differentiation.

$\Psi = C e^{\alpha x} + D e^{-\alpha x}$

The constant C is very small, so the increasing wave function is negligible. The decaying function is the dominant solution and it shows small probability of coming out at the other end.

The rest of the game is similar where you equate the wave functions and the first derivative in all three regions and find the value of constants in terms of k and α. Then you calculate transmission and reflection probabilities.

The bottom line is that the transmission probability is not zero. The particle can still come out at the other end. This is astonishing. It is like saying that if you come across a mountain and you do not have the energy to climb it, do not worry. Keep pushing against the mountain, there is a small chance that you can tunnel through the mountain.

Quantum tunneling seems all airy fairy, but it really happens. The radioactive decay uses quantum tunneling. The alpha particle emitted by uranium atom is an example of this phenomenon. The alpha particle is positively charged and is repulsed by the positively charged nucleus. But the alpha particle has to overcome the barrier of strong nuclear force that binds the nuclear particles. Despite the barrier being too high for the kinetic energy it carries; the alpha particle still tunnels through. The probability of tunneling is very small but alpha particle keeps on bombarding against the barrier and eventually it tunnels through. The moral of the story is never give up, anything is possible!

Quantum Harmonic Oscillator

The modeling of harmonic oscillator tells us a lot about real world quantum systems. The atoms vibrate around one another bound by potential energy of their electrical attraction or repulsion. So, we need to study a simple model and see what solutions we get.

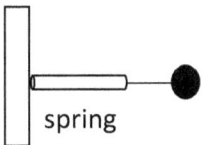
spring

If you pull on the ball, the spring gets stretched and the force acts in the opposite direction to the movement of the ball.

The restoring force is given by $F = -kx$

The potential energy is given by $\int F dx = \int -kx\,dx = \frac{1}{2}kx^2$

k is the spring constant and its frequency of vibration is given by $\omega = \sqrt{\frac{k}{m}}$

The potential energy is $\frac{1}{2}m^2\omega^2 x^2$

Our mission is to solve the Schrodinger equation

$$H\Psi = E\Psi$$

K. E $= \frac{1}{2}mv^2 = \frac{p^2}{2m} = -\frac{\hbar^2}{2m}\frac{d^2\psi}{dx^2}$

$$-\frac{\hbar^2}{2m}\frac{d^2\psi}{dx^2} + \frac{1}{2}m^2\omega^2 x^2 = E\Psi$$

There are too many constants and units in the equation. It needs change of variables.

$$k = \frac{2E}{\hbar\omega}$$

Note energy /energy units cancel to make it dimensionless

$$y = \sqrt{\frac{mw}{\hbar}}x$$

Here distance/distance units cancel to make it dimensionless

Our equation becomes

$$\frac{d^2\psi}{dy^2} = (y^2 - k)\Psi$$

The solution is easy in extreme conditions. If y is very large, approaching infinity, the equation becomes

$$\frac{d^2\psi}{dy^2} = y^2\Psi$$

The solution to the above equation is again exponentials

$$\psi = c_1 e^{-\frac{y^2}{2}} + c_2 e^{+\frac{y^2}{2}}$$

$e^{+\frac{y^2}{2}}$ is no good as it blows up at large x. We do not want probability to keep increasing as you go farther and farther from the spring.

You can check this is the right solution by putting it back in the equation.

$\frac{d^2 e^{-\frac{y^2}{2}}}{dy^2}$ calculation is needed

First derivative is $\frac{de^{-\frac{y^2}{2}}}{dy} = -ye^{-\frac{y^2}{2}}$

Second derivative is $-\frac{dy}{dy}e^{-\frac{y^2}{2}} + (-y)\frac{de^{-\frac{y^2}{2}}}{dy} = (-1+y^2)e^{-\frac{y^2}{2}}$

Since y^2 is a very big number, of order $\frac{mw}{\hbar}$, we can ignore -1.

We get $\dfrac{d^2 e^{-\frac{y^2}{2}}}{dy^2} = y^2 e^{-\frac{y^2}{2}}$

Point proven. If you want to look cool, you write Q.E.D., Latin for quod erat demonstrandum meaning what was to be demonstrated. The solution in extreme cases is called asymptote analysis.

So far, our solution is Ψ = remaining solution(s) $e^{-\frac{y^2}{2}}$

Let's put back the solution into the Schrodinger equation.

After few steps of product differentiation

$$\dfrac{d^2 \psi}{dy^2} = \left(\dfrac{d^2 s}{dy^2} - 2y\dfrac{ds}{dy} - ys\right) e^{-\frac{y^2}{2}}$$

The other side of the equation is $(y^2 - k)\Psi$ or $(y^2 - k)s\, e^{-\frac{y^2}{2}}$

Putting it all together

$$\dfrac{d^2 s}{dy^2} - 2y\dfrac{ds}{dy} + (k-1)s = 0$$

The next step is the power series method. The house that this equation builds is made up of bricks or pieces that are unique. They fit onto each other exactly and each brick helps to build the next one and so on.

$s = c_0 + c_1 y + c_2 y^2 \ldots$

This is like drawing painting of Mona Lisa, start with one paint stroke, keep on drawing till you get the final painting.

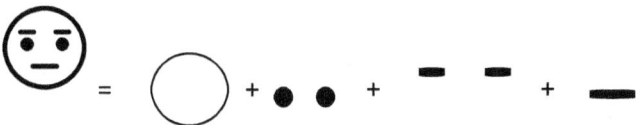

That's your power series!

$$s = \sum_{j=0}^{\infty} c_j y^j$$

$$\frac{ds}{dy} = \sum_{j=0}^{\infty} j c_j y^{j-1}$$

$$\frac{d^2 s}{dy^2} = \sum_{j=0}^{\infty} j(j-1) c_j y^{j-2}$$, you can add +2 to everything to make exponent y^j

We need some manipulation to have exponents of y to be y^j for ease of calculations, which will result in our equation becoming

$$\sum_{j=0}^{\infty} [(j+1)(j+2)\, c_{j+2} - 2j c_j + (k-1) c_j]\, y^j = 0$$

For this equation to be true, all coefficients have to cancel each other. This is only if

$$c_{j+2} = \frac{(2j+1-k)}{(j+1)(j+2)} c_j$$

This is the recursion formula to build the pieces of the solution. There will be odd and even series if you put numbers into j.

How high can j go?

If j is a very big number, then other terms can be ignored, and formula will b

$$c_{j+2} = \frac{2j}{j^2} c_j = \frac{2 c_j}{j}$$

The odd terms will look like the following

$c_{1+2} \approx \frac{2}{1} c_1$ or $c_3 = 2c_1$

$c_{3+2} \approx \frac{2}{3} c_3 = \frac{2}{3} \times 2c_1$

$c_{5+2} \approx \frac{2}{5} c_5 = \frac{2}{5} \times \frac{2}{3} \times 2c_1$

Do you see a pattern, its $\frac{c}{\left(\frac{j}{2}!\right)}$

The solution $s = \sum_{j=0}^{\infty} \frac{c}{\left(\frac{j}{2}!\right)} y^j = c \sum \frac{y^{2j}}{j!}$

This type of expression is called power series. It is a well-known function of Taylor series extension where $\sum \frac{x^n}{n!} = e^x$

$s = ce^{y^2}$

This is not a good solution as it will blow up at large y. So, we need to put the limit on j where the series must terminate.

$c_{jmax+2} = \frac{(2jmax+1-k)}{(jmax+1)(jmax+2)} c_{jmax} = 0$

This is true if $(2j_{max} + 1 - k) = 0$

Let's call $j_{max} = n$

$2n + 1 = k$

Now $k = \frac{2E}{\hbar \omega}$

$E = \left(n + \frac{1}{2}\right) \hbar \omega$

n=0,1,2,3...

The energy cannot take any value but its quantized. When you put a limit on something, waves have to fit in that limit e.g. on a circle, only certain number of waves can be fitted, either full(integer) or none not 0.7 or 0.6 of a wave!

Note the series termination we did was for odd series. You have to do separate requirement for the even series. But since the potential is symmetrical, the same procedure works in the even series as well. Think of odd and even series as the oscillator moving right or left of the origin. The right and left or odd and even series are symmetrical to each other.

As you put values of j and calculate the wave function, you will get polynomials. These are called Hermite polynomials. There is a table that you can refer to avoid calculations.

e.g. $H_0 = 1, H_1 = 2y, H_3 = 4y^2 - 2$ and so on.

The final solution is

$$\Psi_n = CH_n e^{-\frac{y^2}{2}}$$

Where C is a normalization constant. The purpose of the constant is that when we calculate the probability which is $|\Psi|^2$, the result is 1 and factors cancel out. The particle has to be found somewhere.

For completion sake the constant is of the form $\left(\frac{mw}{\pi\hbar}\right)^{\frac{1}{4}} \frac{1}{\sqrt{2^n n!}}$

What did we learn?

We learned how to solve differential equations. But that's not interesting. How quantum harmonic oscillator differs from the classical one? This will come clear when we look at the probability density. Where can we find the particle?

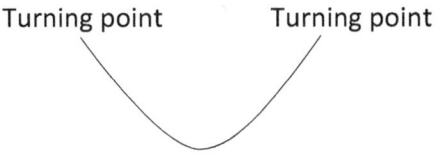

The classic oscillator function is like a parabola where particle moves back and forth between turning points. The particle is at rest at the turning point. Thus, you will find that the probability of finding the particle highest at turning points.

What about the quantum harmonic oscillator?

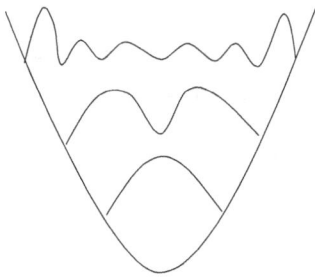

The quantum particle in the ground state is more likely to be found at the center not at the turning points. The next excited state has nodes where there is forbidden zone at the center where particle will never be found. As you go into higher excited states, the nodes keep increasing. When n is very high, then the probability resembles the classical harmonic oscillator.

Alternate Method

There is a different way to solve the Schrodinger equation. Pure algebraic means can sometimes be cleverly used to solve the equation. Quantum harmonic oscillator is a perfect model to use the algebraic method. The method is relatively easy and more elegant to use. However, mathematical difficulty is in the mind of the beholder.

$$H = \frac{p^2}{2m} + \frac{1}{2}m^2\omega^2 x^2$$

Factorization can make things easier.

$$a^2 + b^2 = (a + ib)(a - ib)$$

Unfortunately, we cannot factorize the Hamiltonian this way as x and p are not numbers but operators. The operators do the measurement.

Measuring p then measuring x ≠ measuring x then measuring p

This is uncertainty principle.

The order of operators is important especially if they do not commute.

Doing commutation expresses the uncertainty mathematically

$[x, p]$ = measure x measure p - measure p measure x

If both quantities can be measured simultaneously and do not affect each other then we get zero as the answer.

But we already know that position and momentum cannot be simultaneously known, and we derived this relation earlier

$$[x, p] = i\hbar$$

The name of the game is keep commuting and substituting!

Some of this is repetition as we did commutators in the earlier chapter, but I am deliberately trying to repeat important points so that it sinks in.

We invent ladder operators that we hope will factorize the Hamiltonian.

The trick is to make them dimensionless. We do not have to worry about keeping track of units.

The length dimension = $\frac{m\omega}{\hbar}$

Dimensionless= $\sqrt{\frac{m\omega}{2\hbar}} x$

The momentum dimension= $m\hbar\omega$

Dimensionless=$i \frac{p}{\sqrt{2m\hbar\omega}}$

These are our operators

$A^- = \sqrt{\frac{m\omega}{2\hbar}} x + i \frac{p}{\sqrt{2m\hbar\omega}}$

$A^+ = \sqrt{\frac{m\omega}{2\hbar}} x - i \frac{p}{\sqrt{2m\hbar\omega}}$

What to do next?

Multiply and commute.

$A^+ A^- = (\sqrt{\frac{m\omega}{2\hbar}} x - i \frac{p}{\sqrt{2m\hbar\omega}}) (\sqrt{\frac{m\omega}{2\hbar}} x + i \frac{p}{\sqrt{2m\hbar\omega}})$

Or $\frac{m\omega}{2\hbar}x^2 + \frac{p^2}{2m\hbar\omega} + \frac{i}{2\hbar}(xp - px)$

$xp - px = [x,p] = i\hbar$

Cleaning up the equation and multiplying by $\hbar\omega$, we get

$\hbar\omega\, A^+A^- = \frac{p^2}{2m} + \frac{1}{2}m^2\omega^2 x^2 - \frac{1}{2}\hbar\omega$

We got our Hamiltonian back

$H = \hbar\omega(A^+A^- + \frac{1}{2})$

Let's commute the operators

$[A^-, A^+] = A^-A^+ - A^+A^-$

It will simplify to

$\approx [x,x] + [p,p] + \frac{i}{\hbar}[x,p]$

$= 0 \;+\; 0 \;+ \frac{i}{\hbar} \times i\hbar = 1$

Similarly, $[A^+, A^-] = -1$

What's left, commute the operators with the Hamiltonian.

$[H, A^-] = [\,\hbar\omega(A^+A^- + \frac{1}{2}), A^-\,]$

We do not have to worry about ½ as it's just a number and commutes with A^-.

$\hbar\omega[A^+A^-, A^-] = \hbar\omega A^-[A^+, A^-] = \hbar\omega A^- \times \text{-}1 = -\hbar\omega A^-$

Similarly, $[H, A^+] = +\hbar\omega A^+$

What's the point of all this?

Now we apply these relations to find the eigenvalues of the Schrodinger equation.

First let's apply the commutation relations on the wave function

$[H, A^-]\Psi = -\hbar\omega A^-\Psi$

$HA^-\Psi - A^-H\Psi = -\hbar\omega A^-\Psi$

$A^-H\Psi$ term can be simplified by using $H\Psi = E\Psi$

$A^-H\Psi = A^-E\Psi$ and E is just a number, it can be moved around unlike the operators where order is important

$HA^-\Psi - A^-E\Psi = -\hbar\omega A^-\Psi$

$H(A^-\Psi) = (E - \hbar\omega)A^-\Psi$

Similarly, $H(A^+\Psi) = (E + \hbar\omega)A^+\Psi$

We just found that operators are eigenvectors of Hamiltonian. In plain English it means that once the operator acts on the wave function and you then measure energy by the action of Hamiltonian, you will find the energy value is either increased or decreased by $\hbar\omega$ units.

That's why the operators are called ladder operators. A^+ is the raising operator that increases the energy of a quantum state by $\hbar\omega$. A^- is the lowering operator that decreases the energy of a quantum state by $\hbar\omega$.

Note the ladder operators are not Hermitian, meaning they cannot be used to do real measurements. The eigenvalues are not real as well. They are just for having fun! They tell us how far apart the energy states of the oscillator are.

The lowering operator cannot keep lower the energy. There has to be the lowest energy state that cannot be lowered further.

$A^-\Psi_{min} = 0$

$H \Psi_{min} = \hbar\omega(A^+A^- + \frac{1}{2}) \Psi_{min}$

$= \hbar\omega A^+A^- \Psi_{min} + \frac{1}{2} \hbar\omega \Psi_{min}$

Since $A^- \Psi_{min} = 0$

$H \Psi_{min} = \frac{1}{2} \hbar\omega \Psi_{min}$

$\frac{1}{2} \hbar\omega$ is the lowest energy or the ground state.

If we apply the raising operator, the first excited state energy is $\frac{1}{2} + 1 = \frac{3}{2} \hbar\omega$

This means $E_n = \left(n + \frac{1}{2}\right) \hbar\omega$

The energy is quantized which means it can occupy only certain values.

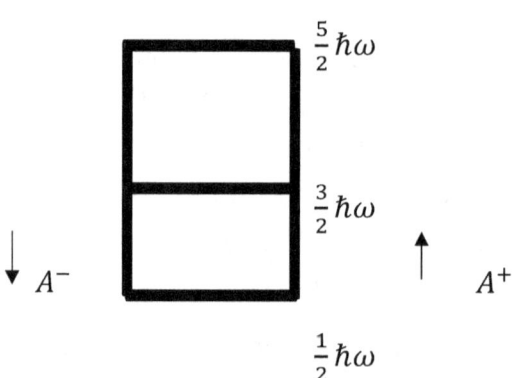

The eigenfunction of the ground state is easy to find.

$A^- \Psi_{min} = 0$

$$\left(\sqrt{\frac{m\omega}{2\hbar}}x + i\frac{p}{\sqrt{2m\hbar\omega}}\right)\Psi_{min} = 0$$

Multiplying both sides by $\sqrt{2m\hbar\omega}$

$$(m\omega x + ip)\Psi_{min} = 0$$

$$\left(m\omega x + i\left(-i\hbar\frac{d}{dx}\right)\right)\Psi_{min} = 0$$

$$\frac{d\Psi_{min}}{dx} = -\frac{m\omega x}{\hbar}\Psi_{min}$$

The solution to the above equation is $\Psi_{min} = C\, e^{\frac{-m\omega x^2}{2\hbar}}$

We have found the eigen functions and eigenvalues of Harmonic oscillator, job done.

$$H|\Psi_{n-1}\rangle = E_{n-1}|\Psi_{n-1}\rangle = (E_n - \hbar\omega)|\Psi_{n-1}\rangle$$

$$H|\Psi_{n+1}\rangle = E_{n+1}|\Psi_{n+1}\rangle = (E_n + \hbar\omega)|\Psi_{n+1}\rangle$$

e.g. $H|\Psi_4\rangle = E_4|\Psi_4\rangle = (E_5 - \hbar\omega)|\Psi_4\rangle$

This equation simply states that when Hamiltonian acts on the fourth excited state, you get the energy of the fourth excited state which is one unit of $\hbar\omega$ less than the fifth excited state.

Chapter 5

Hydrogen Atom

The solution of Hydrogen atom by quantum mechanics is a spectacular achievement of this theory. Quantum mechanics gives very accurate experimental results, which give credence to the fundamentals of the theory. Hydrogen is the simplest atom and available abundantly in the universe. It is no small feat to explain the structure of hydrogen atom with such detail and accuracy. When we say solution of hydrogen atom, we are referring to the energy levels and probability of finding the electron around nucleus etc.

Where do we start?

The standard recipe is the Schrödinger equation. It tells us the value of measurements, which is what we are looking for. To simplify, we assume the potential is time independent so that we can use time independent Schrödinger equation.

$H\Psi = E\Psi$

Hamiltonian is the operator (measuring device), acting on the thing to be measured (wave function) and the energy levels are the results of the measurement.

$\Psi = \sum c_\varepsilon \Psi_\varepsilon$

When we solve the hydrogen atom, we will have the whole collection of the probabilities (c_ε) of each wavefunction with their corresponding energy level.

The Hamiltonian is made of kinetic and potential energies

$$\frac{p^2}{2m} + V$$

The hydrogen atom is a real particle and it lives in three dimensions. So, we need to add KE for all three directions.

$$\frac{p_x^2}{2m} + \frac{p_y^2}{2m} + \frac{p_y^2}{2m}$$

The operator for momentum p_x itself is $-i\hbar \frac{\partial}{\partial x}$ or $-i\hbar \nabla$ where del $\nabla = \frac{\partial}{\partial x}$

$$p_x^2 = (-i\hbar \nabla)^2$$

Similarly, we add operator for the y and z direction

$$p_y^2 = (-i\hbar \nabla)^2 \text{ and } p_z^2 = (-i\hbar \nabla)^2$$

$$\nabla^2 = \frac{\partial^2}{\partial x^2} + \frac{\partial^2}{\partial y^2} + \frac{\partial^2}{\partial z}$$

The Schrödinger equation becomes

$$\left(\frac{-\hbar^2}{2m}\nabla^2 + V\right)\Psi = E\Psi$$

$$\left(\frac{-\hbar^2}{2m}\nabla^2\right)\Psi + (V - E)\Psi = 0$$

We need to use spherical coordinates as they are easier to work with

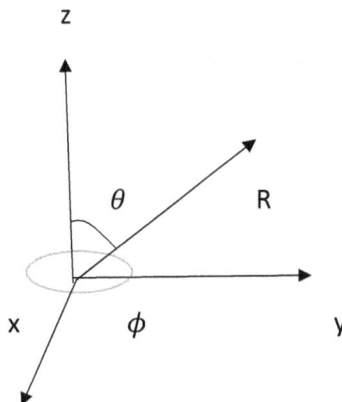

The cartesian coordinates can be converted to the spherical coordinates with basic algebra and trigonometry

$x = r \sin\theta \cos\phi$

$y = r \sin\theta \sin\phi$

$z = r \cos\theta$

The spherical coordinates have the following range

r= 0 → ∞

$\theta = 0 \to \pi$

$\phi = 0 \to 2\pi$

Converting to spherical coordinates causes ∇^2 to become

$$\frac{1}{r^2}\left[\frac{\partial}{\partial r}\left(r^2 \frac{\partial}{\partial r}\right) + \csc\theta \frac{\partial}{\partial \theta}\left(\sin\theta \frac{\partial}{\partial \theta}\right) + \csc^2\theta \frac{\partial^2}{\partial \phi^2}\right]$$

Putting it into Schrodinger equation

$$\left(\frac{-\hbar^2}{2m}\frac{1}{r^2}\left[\frac{\partial}{\partial r}\left(r^2 \frac{\partial}{\partial r}\right) + \csc\theta \frac{\partial}{\partial \theta}\left(\sin\theta \frac{\partial}{\partial \theta}\right) + \csc^2\theta \frac{\partial^2}{\partial \phi^2}\right]\right)\Psi + \left(V - E\right)\Psi = 0$$

This looks like a formidable equation. How do we tackle it?

We use the time-tested colonial policy of "Divide and Rule".

The solution of Ψ can be separated into three parts

$$\Psi = R(r)\Theta(\theta)\Phi(\phi)$$

We have divided solution into radial distance part and two angular parts (θ & ϕ).

Replacing Ψ in the Schrödinger equation with separate radial and angular solutions and doing some manipulation will result in the following

$$\frac{1}{R}\frac{d}{dr}\left(\frac{r^2 dR}{dr}\right) - \frac{2mr^2}{\hbar^2}(E-U) + \frac{1}{\Theta}\csc\theta \frac{\partial}{\partial\theta}\left(\sin\theta \frac{\partial\Theta}{\partial\theta}\right) + \frac{1}{\Phi}\csc^2\theta \frac{\partial^2\Phi}{\partial\phi^2} = 0$$

Note some books write $\frac{\partial(\theta\phi)}{\partial\theta}$ but $\frac{\partial\phi}{\partial\theta} = 0$ so it's better to leave those terms out.

You see we have radial and angular terms equal to constant(zero). The radial and angular terms do not depend on each other. This means radial and angular terms are equal to some constant as well.

$[R] + [\theta, \phi] = 0$

R= constant

θ, ϕ= some other constant

Think of it like tea bag + milk = one cup of tea(constant)

This means tea bag= 1(constant)

Milk = one cup(constant)

Only then you can get one cup of tea.

What to do with the constants?

We call constant of $[\theta, \phi] = -l(l+1)$

There is a bit of cheating here as solutions are known and the importance of constants is known as well, that's why they are written in a suggestive way.

$$\frac{1}{\Theta}\csc\theta \frac{\partial}{\partial\theta}\left(\sin\theta \frac{\partial\Theta}{\partial\theta}\right) + \frac{1}{\Phi}\csc^2\theta \frac{\partial^2\Phi}{\partial\phi^2} = -l(l+1)$$

We want to separate θ and ϕ, so multiplying by $sin^2\theta$ will clean up the equation.

Note $\csc\theta = \frac{1}{\sin\theta}$

$$[\frac{1}{\Theta}\sin\theta \frac{d}{d\theta}\left(\sin\theta \frac{d\Theta}{d\theta}\right) + l(l+1)\sin^2\theta] + \frac{1}{\Phi}\frac{d^2\Phi}{d\phi^2} = 0$$

Also note that when we separate variables, partial differentiation $\frac{\partial}{\partial\theta}$ is replaced by ordinary differentiation $\frac{d}{d\theta}$.

Doing further separation of angular parts by a different separation constant

$$\frac{1}{\Theta}\sin\theta \frac{d}{d\theta}\left(\sin\theta \frac{d\Theta}{d\theta}\right) + l(l+1)\sin^2\theta = m^2$$

$$\frac{1}{\Phi}\frac{d^2\Phi}{d\phi^2} = -m^2$$

First, let's solve the ϕ equation as it's the easiest.

The solution is $e^{im\phi}$ or $e^{-im\phi}$

This is basic exponential differentiation, where factors come in front of the exponential keeping the exponential unchanged each time you differentiate.

How do we interpret the solution?

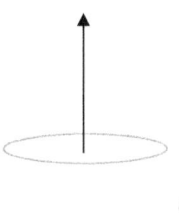

The ϕ angle circles around the z axis

Φ → 2π → Φ

The solution comes back to itself after completing the full circle or 2π.

$$e^{im\phi} = e^{im(\phi+2\pi)}$$

This is true if m is an integer. Obviously if you are moving in a circle, you can reach the same point only if you complete a full circle one time, two times, three times etc. You cannot complete half a circle and reach the same point on the circle!

m ranges from …….-3,2,1,0,1,2,3……

m is called the magnetic quantum number for historical reasons.

It gives the component of angular momentum in a specific direction. Z axis is the traditional preferred direction.

When m is 0, it means angular momentum has no component sticking out in z direction. So, the probability to find particle around z axis is same everywhere. When m is one, there is one full wave fit into the circle around z axis, when m is two, there are two standing waves fit into the circle and so on. When there are waves fit into the circle, the probability will not be smooth, there will be uneven distribution of probability around z axis. Practically speaking, m quantum number determines the orientation of the atomic orbital. It will become clear as we solve the θ angular equation.

Next, let's solve the θ equation

$$sin\theta \frac{d}{d\theta}\left(sin\theta \frac{d\Theta}{d\theta}\right) + [\,l(l+1)sin^2\theta - m^2\,]\Theta = 0$$

Our physicist is stuck, he is not sure how to solve the equation. He goes to the nearby math hardware store and asks for help.

Can you help me build the house of solutions for my equation?

The mathematician immediately recognizes the equation. He says that the great mathematician Legendre has solved a similar equation. It is not in spherical coordinates, but we can work around it.

How do you solve the equation?

The mathematician says, if you want to build a house, you need bricks. The bricks are part of the power series method where bricks come in different shapes and sizes to build the house based on your equation. They have to fit exactly like pieces of a puzzle and each piece influences the next one and so on.

The second order differential equation comes with two type of solutions.

$$y = C_1 P_l^m + C_2 Q_l^m$$

These are called Legendre functions of first and second kind. But based on your spherical coordinates, Q_l^m function will blow up and give unrealistic answers. So, we need only functions of the first kind. Since m and l are integers, the functions are called Legendre polynomials.

$$y = \Sigma c_n x^n$$

These are your bricks and after putting in the power series solution into the differential equation, you get the recursion formula for Legendre polynomials which gives a quick way to build the solutions. There are several steps involved in getting to recursion formula which are not necessary for a physicist to know. Learn how to use mathematical tools, not how they are made says the mathematician smugly.

A compact way to write the brick forming formula in this case is called Rodrigues formula.

$$P_l = \frac{1}{2^l l!} \left(\frac{d}{dx}\right)^l (x^2 - 1)^l$$

There is a table of Legendre polynomials (for $l = 0,1,2.3 \ldots$)where you have the catalogue to choose the size of the bricks for your house.

e.g. $l = 1$, $P_1 = \frac{1}{2^1 1!} \left(\frac{d}{dx}\right)^1 (x^2 - 1)^1 = x$

The l cannot be negative for the formula to work.

Where is m in this?

The mathematician says the ordinary Legendre equation works for m=0 only.

That's no good, m should be able to take integer values.

The mathematician says that we will have to use the associated Legendre equation for non-zero m. Moreover, replace $sin\theta, sin^2\theta$ in the equation with $cos\theta$ (recall $sin^2\theta + cos^2\theta = 1$). Then set $cos\theta = x$, we will get the associated Legendre equation from θ equation

$$\frac{d}{dx}(1-x^2)\frac{d}{dx}P_l^m + [l(l+1) - \frac{m^2}{1-x^2}]P_l^m = 0$$

The solution is y or $\Theta = CP_l^m(\cos\theta)$

P_l^m are associated Legendre polynomials which serve as the bricks to form our house of angular function.

$$P_l^m = (1-x^2)^{\frac{|m|}{2}}\left(\frac{d}{dx}\right)^{|m|}P_l$$

The associated Legendre polynomials do contain Legendre polynomials P_l.

This formula restricts m to be from $-l$ to $+l$. The solution is zero if m>l.

e.g. if $l = 1, m = 2$

$$P_1^2 = (1-x^2)^{\frac{|2|}{2}}\left(\frac{d}{dx}\right)^2 x = (1-x^2)\frac{d}{dx}1 = 0$$

So, it only works if $l > m$

$$P_1^0 = (1-x^2)^{\frac{|0|}{2}}\left(\frac{d}{dx}\right)^0 x = x = \cos\theta$$

There is a table of associated Legendre polynomials as well.

The angular solutions are combined together to give a joint solution

$$Y = \Theta(\theta)\Phi(\phi)$$

The combined solutions are called Spherical Harmonics. There is another table of spherical harmonics based on the values of l and m.

A reminder that the probability of finding a particle is not Y but $|Y|^2$.

What have we learned from the angular solutions?

The l is the angular momentum quantum number with m quantum number representing its projection on z axis. The probabilities of finding particles look like

$l = 0, m = 0$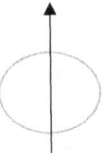

The probability of finding an electron is same everywhere around the nucleus. The radius of the probability cloud will be determined by the radial equation. This is very different than the classical picture of electron moving around the nucleus like earth revolves around the sun. The probability is a standing wave spread around the nucleus. It is not changing with time. When you measure the electron, it will randomly appear anywhere in the cloud. You will find the electron at one spot only at a time due to collapse of the probability cloud on measurement. This is the s orbital in chemistry notation.

$l = 1, m = 0$

There is some angular momentum here, the lobes tell us that electron is pushed out due to angular momentum. The antinode at the origin tells us that electron cannot be found there. This is all wave theory, where probability is enhanced by constructive interference and reduced by destructive interference at the antinodes. The probability around z axis is same as m=0. Here we have p_z orbital.

$l = 1, m = +1 \text{ or } -1$

Here the lobes are oriented along x or y axis. The probability along z axis is uneven as it should be as m is non-zero. As you can see m determines orientation of the atomic orbitals.

$l = 3, m = 2$

As the angular momentum increases, you will find more lobes and there is spreading out of the probability cloud. This is expected as higher angular momentum tends to push electron out wards. This is an example of d orbital.

Radial equation

Last but not the least is the radial equation that needs to be solved.

$$\frac{1}{R}\frac{d}{dr}\left(\frac{r^2 dR}{dr}\right) - \frac{2mr^2}{\hbar^2}(V-E) = l(l+1)$$

We choose $l(l+1)$ as the separation constant here.

$$\frac{d}{dr}\left(\frac{r^2 dR}{dr}\right) - \frac{2mr^2}{\hbar^2}(V-E)R = l(l+1)R$$

This does not look like a Schrodinger equation, so we need to change some variables.

Let $u = rR$ and do basic product differentiation.

$$\frac{d}{dr}\left(\frac{r^2 dR}{dr}\right) = \frac{d}{dr}r^2\left(\frac{d}{dr}\right)\frac{u}{r} = \frac{d}{dr}r^2\left(-\frac{1}{r^2}u + \frac{1}{r}\frac{du}{dr}\right)$$

Or $\frac{d}{dr}(-u + r\frac{du}{dr}) = \frac{-du}{dr} + \frac{du}{dr} + \frac{rd^2u}{dr^2} = \frac{rd^2u}{dr^2}$

So, we converted two derivates into a single derivative, this is better. Doing further manipulation will cause the radial equation to look like Schrodinger equation.

$$\frac{-\hbar^2}{2m}\frac{d^2u}{dr^2} + [V + \frac{\hbar^2}{2m}\frac{l(l+1)}{r^2}]u = Eu$$

The potential energy in a hydrogen atom comes from the electric potential energy between proton and electron and is given by classical expression

$$V = -\frac{1}{4\pi\varepsilon_0}\frac{e^2}{r}$$

The negative sign is a convention and indicates the attractive nature of the potential energy which is taken to be zero at infinity and gets negative as charges come closer. You have to do positive work to remove the charges.

$\frac{\hbar^2}{2m}\frac{l(l+1)}{r^2}$ is an extra term and it denotes centrifugal force trying to push electron out of the orbit. As you can see if $l=0$ or if there is no angular momentum, there is no centrifugal force.

$$\frac{-\hbar^2}{2m}\frac{d^2u}{dr^2} + [-\frac{1}{4\pi\varepsilon_0}\frac{e^2}{r} + \frac{\hbar^2}{2m}\frac{l(l+1)}{r^2}]u = Eu$$

The physicist is again stuck. He asks the mathematician.

Make a guess says the mathematician.

Power series could work.

That is correct but not so fast, says the mathematician. If you apply power series method directly, you will get multiple recursion formulae which are not easy to deal with. The prudent approach is to peel of the solutions at the extreme values and then apply the power series formula to the remaining bit.

First, take the case where $r \to \infty$. The terms with r in the denominator can be ignored, so the equation becomes

$$\frac{d^2u}{dr^2} = -\frac{2mE}{\hbar^2}u$$

Let $\alpha = \sqrt{-\frac{2mE}{\hbar^2}}$

We are dealing with bound states here, meaning electrons that are bound in the hydrogen atom, not free. This makes E negative and α real.

The solution to above equation is straightforward.

$$u = Ae^{-\alpha r} + Be^{\alpha r}$$

$Be^{\alpha r}$ is rejected as it blows up meaning probability keep on increasing at large distance which is not realistic.

$$u = Ae^{-\alpha r}$$

Second, take the case where $r \to 0$. The term with r^2 in the denominator is the dominant term as $\frac{1}{(r \to 0)^2}$ approaches infinity so we can ignore other r terms.

$$\frac{d^2u}{dr^2} = \frac{l(l+1)}{r^2}u$$

Here we make a guess that u is made of a polynomial $u = cr^j$

Doing differentiation will lead to

$$\frac{d^2u}{dr^2} = cj(j-1)r^{j-2}$$

This means $cj(j-1)r^{j-2} = \frac{l(l+1)}{r^2}cr^j$

$j(j-1)r^{j-2} = l(l+1)r^{j-2}$

$$j(j-1) = l(l+1)$$

$$j = -l, j = l+1$$

$$u = Cr^{-l} + Dr^{l+1}$$

As $r \to 0$, Cr^{-l} will blow up.

$$u \sim Dr^{l+1}$$

Total solution $= e^{-\alpha r} r^{l+1} v$

Where v is the remaining solution.

What to do with the remaining unsolved solution?

The power series method can be applied to it. There is change of variables required. The change of variables is a standard recipe to simplify differential equations. The variables chosen are usually dimensionless. The reason is that dimensions or units sometimes introduce unnecessary factors in the equations. Moreover, units are based on standard reference units that may not be relevant to system at hand. We have SI units where kilogram, meters etc. are the standard units as they are relevant to humans in daily life. But these units are not that relevant for electron in a hydrogen atom. We want to get rid of the units and remove unnecessary factors from equations and focus on relevant quantities like how things are related.

The term αr can be soaked up by introducing a dimensionless variable ρ.

$$\rho = \alpha r$$

Note $\alpha = \sqrt{-\frac{2mE}{\hbar^2}}$ has units of inverse length.

The factor $\frac{\alpha}{4\pi\varepsilon_0} e^2$ can also be soaked. There are different versions in various books but it's not relevant. The important point is that it has energy stored in the variable. Let's call this variable λ.

We need to put solution $e^{-\alpha r} r^{l+1} v$ into the radial equation

$$\frac{-\hbar^2}{2m}\frac{d^2u}{dr^2} + \left[-\frac{1}{4\pi\varepsilon_0}\frac{e^2}{r} + \frac{\hbar^2}{2m}\frac{l(l+1)}{r^2}\right]u = Eu$$

After doing the change of variables, the equation is changed to find v instead of u. Also, its customary to differentiate with respect to dimensionless ρ rather than r.

I will not go into every step of changing the variables, but the equation will resemble a well-known mathematical equation called the associated Laguerre equation

$$\frac{x\,d^2y}{dx^2} + (y+1-x)\frac{dy}{dx} + \lambda y = 0$$

Which will resemble radial equation after doing variable change

$$\frac{x\,d^2v}{dx^2} + (2(l+1)-x)\frac{dv}{dx} + \lambda v = 0$$

The solution can be found by power series where y or $v = \sum c_i \rho^i$

The detailed solution where you put this power series solution into the equation and find the recursion formula is couple of pages of algebra which I will skip. I hope you have got the idea because we did the same steps while solving harmonic oscillator equation.

The important thing is the recursion formula which gives us the bricks to make the house of radial equation.

$$c_{i+1} = \frac{2(i+l+1) - \lambda(energy\ factor)}{(i+1)(i+2l+2)} c_i$$

If i is a large number, other factors can be ignored, and the relation reduces to

$$c_{i+1} \approx \frac{2}{i} c_i$$

The terms will look like the following

$$c_{1+1} \approx \frac{2}{1} c_1 \text{ or } c_2 = 2c_1$$

$$c_{2+1} \approx \frac{2}{2} c_2 = \frac{2}{2} \times 2c_1$$

$$c_{3+1} \approx \frac{2}{3} c_3 = \frac{2}{3} \times \frac{2}{2} \times 2c_1$$

$$c_{4+1} \approx \frac{2}{4} c_4 = \frac{2}{4} \times \frac{2}{3} \times \frac{2}{2} \times 2c_1$$

Do you see a pattern, its $\frac{2^i}{i!}$

The solution $v = \sum c_i \rho^i = c_0 \sum \frac{2^i}{i!} \rho^i$

This type of expression is called power series. It is a well-known function of Taylor series extension where $\sum \frac{x^n}{n!} = e^x$

$$v = c_0 e^{2i}$$

This is not a good solution as it will blow up at i or large distance away from nucleus. So, we need to put a limit to i as we do not want our hydrogen solution to extend forever as electrons are closely bound to hydrogen nucleus and we cannot make hydrogen atom expand too much.

We propose that when i reaches maximum, solution is zero.

$$c_{i(max+1)} = \frac{2(i+l+1) - \lambda(energy\ factor)}{(i+1)(i+2l+2)} c_{imax} = 0$$

This is true if $2(imax + l + 1) - \lambda = 0$

Let's call $imax + l + 1 = n$

This our principal quantum number.

$2n = \lambda(energy\ factor)$

The principal quantum number determines the energy of the orbital. The energy is quantized as n is an integer that goes from 1,2,3……

This expression also establishes relationship between n and l quantum numbers.

$l = n - i - 1$

$l = 0,1,2,3 \ldots n - 1$

e.g. if n= 1 then l= 0. If n=2 then l= 0,1.

The exact expression is $E_n = -\dfrac{me^4}{2\hbar^2 n^2 (4\pi\varepsilon_0)^2}$

Now we also have $\rho = \alpha r$ and $\alpha = \sqrt{-\dfrac{2mE}{\hbar^2}}$

When you combine these equations with the energy equation of principal quantum number, we get

$\alpha = \dfrac{me^2}{n 4\pi\varepsilon_0 \hbar^2}$

α has units of inverse length, so $\dfrac{1}{\alpha}$ has units of length.

The length is $\dfrac{n 4\pi\varepsilon_0 \hbar^2}{me^2}$

For n=1

$$a_0 = \frac{4\pi\varepsilon_0 \hbar^2}{me^2}$$

This is called the Bohr's radius. It gives the approximate distance of electron from the nucleus in the ground state of hydrogen.

The recursion formula for v can be written in a compact form as a polynomial. We saw similar thing in the angular equation where recursion formula was packaged as associated Legendre polynomial that make up the solution for associated Legendre equation. Similarly, associated Laguerre polynomials form the solutions of the associated Laguerre equation.

$$v = L_{n-l-1}^{2l+1}(\text{where } 2\rho = x)$$

If you set $k = 2l + 1$ and $j = n - l - 1$ then associated polynomials can be calculated by formula

$$L_j^k = (-1)^k \frac{d^k}{dx^k} L_{j+k}$$

L_{j+k} itself is a Laguerre polynomial given by formula by letting $j + k = q$

$$L_q = e^x \frac{d^q}{dx^q} e^{-x} x^q$$

I know this is a mess. Just don't forget the basic facts. We are building a house (solution of v) which is made of bricks (associated Laguerre polynomials). The bricks are further made up of further ingredients (Laguerre polynomials).

There is table of these polynomials available.

We have found the radial solution after much hard work.

$$u = e^{-\rho} \rho^{l+1} L_{n-l-1}^{2l+1}(2\rho)$$

We know that $\rho = \alpha r$ and $\frac{1}{\alpha} = a_0$ (Bohr's radius) so

$$u = e^{-\frac{r}{na_0}} \left(\frac{2r}{na_0}\right)^{l+1} L_{n-l-1}^{2l+1}\left(\frac{2r}{na_0}\right)$$

Further $u = rR$

$$R = (\text{Constant})\, e^{-\frac{r}{na_0}} \left(\frac{2r}{na_0}\right)^{l} L_{n-l-1}^{2l+1}\left(\frac{2r}{na_0}\right)$$

Final solution of Hydrogen wave function

$$\Psi = R(r)\Theta(\theta)\Phi(\phi) = \text{Normalization constant}\; e^{-\frac{r}{na_0}} \left(\frac{2r}{na_0}\right)^{l} L_{n-l-1}^{2l+1}\left(\frac{2r}{na_0}\right) Y_l^m(\theta,\phi)$$

By putting different values of n, l and m, we can generate all hydrogen energy eigenfunctions e.g. ψ_{100} is the notation for ground energy eigenfunction.

Physicist thanks the mathematician for the help.

"Mathematics trumps Physics" says the mathematician smugly.

Physicist would have none of it. He smilingly says, "PGA championship is won by the person who knows how to use golf clubs, not by the one who designs it!"

Are we done yet?

Not quite. To get the probability density, we have to square the wavefunction $|\psi|^2 dV$. The shapes are not too different than what we got with angular functions alone. We will skip the details of individual wave function density plots.

In summary, we solved the time independent Schrodinger equation for Hydrogen.

$$H\Psi = E\Psi$$

We now know Ψ_{nlm} and their corresponding energy values. Since energy depends only on the n quantum number, there is degeneracy for different values of l quantum number.

$\psi_{211}, \psi_{200}, \psi_{210}$ states have the same energy.

What's the use of finding the energy levels?

The study of Hydrogen energy spectrum is very important. It is like DNA of atoms. The presence of Hydrogen in the universe is detected by looking at the energy spectrum. The correct prediction of energy spectrum by quantum mechanics is a validation of the robustness of the theory.

The energy eigenstates are stationary, but environment does not let them stay that way. The higher energy electron can come down the energy level by emitting a photon. The lower energy electron can also go to higher level by absorbing a photon. The photons emitted are of different energies. Their wavelengths are different which leads to characteristic spectral lines.

The energy of a photon is $E = \hbar\omega$ where ω is angular frequency.

$$E_n = -\frac{me^4}{2\hbar^2 n^2 (4\pi\varepsilon_0)^2}$$

Combining these two, leads to the famous Rydberg formula

$$\omega = \frac{R_y}{\hbar}\left(\frac{1}{n_1^2} - \frac{1}{n_2^2}\right)$$

R_y is Rydberg constant.

The spectral lines have special names e.g. electrons dropping to n=1 form Lyman series, n=2 form Balmer series etc.

Alternate method

Solving differential equations is a tedious business. Physicists always look to bypass solving differential equations. We saw in the case of harmonic oscillator that there is an alternate algebraic way to solve the Schrodinger equation.

$$H\Psi = E\Psi$$

We are interested in eigenvalues and eigenfunctions. The alternate method works for angular momentum only. We are stuck with solving the radial equation the traditional way.

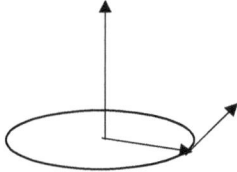

The electron moving around the nucleus possesses angular momentum. It is given by the cross product of radius and momentum, with direction given by right hand rule.

$$L = r \times p = r\, p\, \sin\theta$$

The maximum angular momentum is when r and p are perpendicular as sin 90 is 1. The angular momentum is zero when angle between r and p is zero. This makes sense e.g. if you are trying to rotate a wheel, you have to push it tangentially. If you push or pull on the wheel towards you, the wheel cannot be rotated.

How do we calculate the cross product?

In 3 dimensions, cartesian coordinates i, j and k are perpendicular to each other

$(a_1 i + a_2 j + a_3 k)(b_1 i + b_2 j + b_3 k)$

We have to multiply each term and use the fact that $i \times i = 0$ (sin 0)

And $i \times j = k$ (sin 90)

It is convenient to take the cross product using matrix notation

$$a = \begin{pmatrix} b_x \\ b_y \\ b_z \end{pmatrix} \times \begin{pmatrix} c_x \\ c_y \\ c_z \end{pmatrix} \text{ or}$$

$$a_x = \begin{pmatrix} b_y \\ b_z \end{pmatrix} \times \begin{pmatrix} c_y \\ c_z \end{pmatrix} = b_y c_z - b_z c_y$$

Similarly, a_y and a_z can be calculated.

We have the angular momentum components as

$L_x = y p_z - z p_y$

$L_y = z p_x - x p_z$

$L_z = x p_y - y p_x$

Putting in the momentum operator $p = \dfrac{\hbar}{i} \dfrac{\partial}{\partial x}$

$$L_z = \frac{\hbar}{i}\left(x\frac{\partial}{\partial y} - y\frac{\partial}{\partial x}\right)$$

The other two components can be likewise derived.

We need uncertainty principle to unwind the Schrodinger equation and get the solution.

Doing commutation expresses the uncertainty mathematically

$[x,p]$ = measure x measure p - measure p measure x

If both quantities can be measured simultaneously and do not affect each other then we get zero as the answer.

But we already know that position and momentum cannot be simultaneously known, and we derived this relation earlier

$$[x,p] = i\hbar$$

Using this fact and lot of substitutions is the name of the game here. So, let's begin.

$[L_x, L_y] = L_xL_y - L_yL_x$ can be calculated by substituting the operator version we calculated earlier. If you carry out the calculation, there will be eight terms, some of which cancel out giving the following expression

$$\left(\frac{\hbar}{i}\right)^2 \left(x\frac{\partial}{\partial y} - y\frac{\partial}{\partial x}\right) = i\hbar L_z$$

Likewise, $[L_y, L_z] = i\hbar L_x$ and $[L_z, L_x] = i\hbar L_y$

We cannot know any two components of angular momentum as they do not commute. The uncertainty principle stops us from knowing so.

What is the magnitude of the total angular momentum

$$L^2 = L_x^2 + L_y^2 + L_z^2$$

Does L^2 commutes with the individual components?

$[L^2, L_x] = [L_x^2, L_x] + [L_y^2, L_x] + [L_z^2, L_x]$

$[L_x^2, L_x] = 0$

Next, we use commutator relations $[AB, C] = A[B, C] + [A, C]B$

$[L_y^2, L_x] = L_y[L_y, L_x] + [L_y, L_x]L_y$

Putting in $[L_x, L_y] = i\hbar L_z$ and doing similar expansion with $[L_z^2, L_x]$

We will find all terms will cancel out

$[L^2, L_x] = 0$

So, we can know the total angular momentum and at most one component.

We need to be creative and come up with ladder operators

$L_+ = L_x + iL_y$

$L_- = L_x - iL_y$

They are not Hermitian operators meaning they cannot be used to measure anything real, but they are mathematical functions that help us find the observables.

What do we do next?

Keep commuting!

$[L_z, L_+] = [L_z, L_x] + i[L_z, L_y] = \hbar L_+$

or $L_z L_+ - L_+ L_z = \hbar L_+$

Likewise

$[L_z, L_-] = \hbar L_-$

Let's solve the Schrodinger equation

$L_z \Psi = c\Psi$

We will cleverly choose eigen value $c = m\hbar$

Now we calculate

$L_z(L_+\Psi) = L_+L_z\Psi + \hbar L_+\Psi$ (from $[L_z, L_+]$)

$= L_+ m\hbar \Psi + \hbar L_+\Psi$

Usually you have to keep order of terms but simple constants ($m\hbar$) or numbers can be moved around

$= (m+1)\hbar L_+\Psi$

We found that $L_+\Psi$ is an eigenfunction of L_z. This tell us that when L_+ acts on Ψ and then you try to measure L_z eigenvalues, the resultant measurement is increased by 1 unit.

L_+ is the raising operating that raises L_z eigenvalues by 1 unit.

Similarly, L_- is the lowering operating that lowers L_z eigenvalues by 1 unit.

How high can ladder go?

It is obvious that value of L_z cannot exceed total angular momentum L.

There is an upper limit where $L_+\Psi_{max} = 0$

Likewise, $L_-\Psi_{min} = 0$

We need to calculate L_+L_-

$L_+L_- = (L_x + iL_y)(L_x - iL_y) = L_x^2 - iL_xL_y + iL_yL_x + L_y^2 = L_x^2 + L_y^2 - \hbar L_z$

Or $L_x^2 + L_y^2 = L_+L_- + \hbar L_z$

$L^2 = L_x^2 + L_y^2 + L_z^2 = L_+L_- + \hbar L_z + L_z^2$

Now a final substitution, I know your patience is getting thin but trust me we are near the end of this commuting and substitution game.

$L^2 \Psi_{max} = L_+ L_- \Psi_{max} + \hbar L_z \Psi_{max} + L_z^2 \Psi_{max}$

Let's put the eigenvalues (remember $H\Psi = E\Psi$)

$L_+ L_- \Psi_{max} = 0$ as non-Hermitian operators have no eigenvalues.

$L_z^2 \Psi_{max} = m_{max}^2 \hbar^2$

$\hbar L_z \Psi_{max} = m_{max} \hbar^2$

Substituting the values in the equation

$L^2 \Psi_{max} = (m_{max}^2 \hbar^2 + m_{max} \hbar^2) \Psi_{max}$

$L^2 \Psi_{max} = m_{max} \hbar^2 (m_{max} + 1) \Psi_{max}$

Likewise, $L^2 \Psi_{min} = m_{min} \hbar^2 (m_{min} - 1) \Psi_{min}$

Since ladder is symmetrical, $m_{max}(m_{max} + 1) = m_{min}(m_{min} - 1)$

Or $m_{min} = -m_{max}$

We can call $m_{max} = l$

In summary

$L_z \Psi = m\hbar \Psi$

$L^2 \Psi = \hbar^2 l(l+1)$

l takes values from m_{min} to m_{max}

Or m takes values from $-l$ to $+l$

Note that m and l can be integers 0,1,2,3 or half integers ½, $\frac{3}{2}$ etc.

We found eigenvalues without solving the differential equation by pure algebraic means.

What's left?

The eigenfunctions.

The calculation is straight forward if you convert Cartesian coordinates into spherical coordinates. It involves fair bit of calculation, but procedure is straight forward. I will not show each step in the conversion as there are set formulae to do conversions but will illustrate the basic principle.

$L_z = \frac{\hbar}{i}\left(x\frac{\partial}{\partial y} - y\frac{\partial}{\partial x}\right)$ in spherical coordinates will become $\frac{\hbar}{i}\frac{\partial}{\partial \phi}$

$L_z \Psi = m\hbar \Psi$

$\frac{\hbar}{i}\frac{\partial \Psi}{\partial \phi} = m\hbar \Psi$

Solution to the above equation is $ce^{im\phi}$

This is the same solution we found when we solved the azimuthal equation

$\frac{1}{\Phi}\frac{d^2\Phi}{d\phi^2} = -m^2$ where solution is again $e^{im\phi}$

The solution of algebraic and differential equations match, that's good.

$\Phi \rightarrow 2\pi \rightarrow \Phi$

The solution comes back to itself after completing the full circle or 2π.

$e^{im\phi} = e^{im(\phi + 2\pi)}$

This restricts the m to be an integer only. The spherical coordinates restrict m to be integer only. There was no restriction when we used pure commutation relations without invoking the spherical coordinates. This has profound physical significance. Since angular momentum happens when an electron moves around the nucleus in space, the coordinates restrict the m quantum number. If spherical coordinates are not involved, then m can be half integer.

This is absurd, how can a particle moving not involve space?

Where is it moving? It has to have some spatial coordinates.

This is precisely what happens in electron spin.

It turns out that electron spin does not involve moving in space. Electron is a point particle without any structure. Electron spin is not spin in three-dimensional space but an internal spin. So, no spherical coordinates are involved and thus m quantum number can have half integer values. I know it sounds so goofy but it's true.

The total angular momentum L^2 can be calculated by converting $L_x^2 + L_y^2 + L_z^2$ into spherical coordinates.

$$L^2 = -\hbar^2 [\frac{1}{\sin\theta}\frac{\partial}{\partial\theta}(\sin\theta \frac{\partial}{\partial\theta}) + \frac{1}{\sin^2\theta}\frac{\partial^2}{\partial\phi^2}]$$

This is similar to the angular equation that was solved earlier, and the solutions were spherical harmonics.

We have all that we need, eigenfunctions and eigenvalues through an alternative method which is relatively simple but not easy.

Chapter 6

Spin

Spin is the quintessential quantum phenomenon. It encapsulates the random and weird nature of quantum mechanics. If you can understand quantum spin, you have understood the soul of the subject.

What is spin?

A body rotating around its axis is spinning. If something is rotating, it possesses spin angular momentum, the direction of which is given by right hand rule.

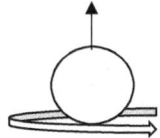

A charged body rotating around its axis also generates a magnetic field or magnetic moment which interacts with an external magnetic field. This helps to detect the magnitude and direction of spin.

How does quantum spin different than classical spin?

First of all, quantum particle like electron is a point particle. There is no structure to it. How can a point rotate on itself?

You may say that we need better resolution to find the structure of electron. This is true for a bigger particle like proton which has structure and made up of quarks. But electron is truly a point particle, based on current evidence. So how a quantum spin is generated, remains a mystery. How electron waves interact at a point to give spin like characteristics is not well understood.

If something is spinning, we can always stop it. We can also make it spin faster or slower. This does not happen with quantum spin. The electron spin is fixed. All electrons have the same spin. It is a fundamental property like electric charge.

The spinning electron is a tiny magnet. It will align or anti-align with a measuring device. We can call it spin up or down, especially if the measuring device is used vertically. If an electron spins like a classic particle, rotating in the anticlockwise direction with the spin vector up then we will find

Measuring device   electron spin up

We put the device vertically in the North-South direction and find the electron spin up. We will say that electron is spinning from right to left and by right hand rule the spin is up. That is expected classically.

Then we tilt the device in the East-West direction, will we measure anything?

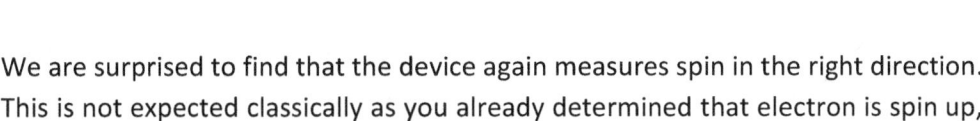

We are surprised to find that the device again measures spin in the right direction. This is not expected classically as you already determined that electron is spin up, why it is now spin right?

In fact, you can tilt the device in any direction, you will always get a reading. It will have either spin aligned, or anti aligned with the device. You can call it spin up or down, left or right, in and out or whatever.

Which is the real direction of the spin in the electron?

This is the real dilemma in quantum mechanics. Nothing is certain. The results are random, and you can only tell the probability of a particular result.

Before the measurement, direction of the electron spin is not fixed. It is in superposition of all directions. Once we measure, superposition collapses and we get a result with certain probability. Let's say we got spin up in the z direction. If we keep on measuring repeatedly in the z direction, we will get spin up only. Otherwise nothing will make sense, experiments are reliable thankfully. But now when we measure spin in the x direction, the superposition of spin right or left collapses and we have 50% chance of getting spin right or left. If we go back and repeat the z direction measurement, we are not guaranteed to get spin up that we got earlier. The chance will be 50% up or down. The moral of the story is that measurement results are random, we can only give chances of a particular result, nothing more.

We cannot find the x, y and z components of spin at the same time. This is uncertainty principle. We can only know one component of the spin and the total spin. The total spin vector cannot be along z or any measured component of spin as the other two components of spin are uncertain. In words of mathematics, spin components are non-commutable.

$[S_x, S_y] \neq [S_y, S_x]$

This is to say when we measure spin in the x direction first then in the y direction, it is not same as measuring spin in the y direction first followed by x direction.

This is the same commutation relation we found with angular momentum too.

$[L_x, L_y] \neq [L_y, L_x]$

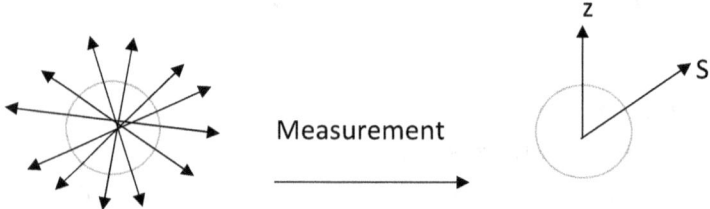

Superposition of all spin orientations   Only total spin and one component is known

We need to go into the nitty and gritty of mathematics of the spin. The good news is lot of the results are similar to angular momentum. So just copy paste formulae of angular momentum.

The total angular momentum of a particle is

$$L^2 = l(l+1)\hbar^2 \Psi_L$$

Likewise, the total spin of a particle is

$$S^2 = s(s+1)\hbar^2 \Psi_s$$

Where s is the spin quantum number. Unlike angular quantum number l, the spin quantum number can have half integer values e.g. $0, \frac{1}{2}, 1, \frac{3}{2}$. The electron has s=1/2.

The z component of the spin angular momentum mimics angular momentum component as well.

$$S_z = m_s \hbar \Psi_s$$

The m_s can have values ranging from -s to +s in integer steps.

The commutation relation of spin is also identical to angular momentum.

$[S_x, S_y] = i\hbar S_z$ and so on for other commutation relations.

Electron spin

The study of the electron spin gives you all you need to know about quantum spin.

In quantum mechanics, solving any problem involves knowing the eigenvectors, eigenvalues and operators.

In other words, we need to know what we are measuring(eigenvectors), what are we measuring with(operators) and what are the measurement results(eigenvalues).

The eigenvalue results are straight forward. All electrons have same spin. The electron is a spin ½ particle. The z component has two values

$+\frac{\hbar}{2}$ or $-\frac{\hbar}{2}$. This is to say $m_s = +\frac{1}{2}$ or $-\frac{1}{2}$ and $s = \frac{1}{2}$.

This is because electrons are tiny magnets, so they have North and South Poles. Since we do not want to confine them to any direction, positive and negative signs are used to represent the poles of the electron magnet.

Note that spin is measured in the units of $\hbar$.

Total electron spin squared $= S^2 = s(s+1)\hbar^2 = \frac{1}{2}\left(\frac{1}{2}+1\right)\hbar^2 = \frac{3}{4}\hbar^2$

$S = \frac{\sqrt{3}}{2}\hbar$

As usual we chose z component as a preferred axis for convenience, but you don't have to. The x and y components too have same eigenvalues of $+\frac{1}{2}$ or $-\frac{1}{2}$.

In fact, you can choose any direction, result is always the same.

So, the problem of eigenvalues is solved.

Now, some notational stuff. The wave function or state vector can be written as

$|s, m\rangle = |\frac{1}{2}, +\frac{1}{2}\rangle$ or $|\frac{1}{2}, -\frac{1}{2}\rangle$

The z component can be written as $|\uparrow\rangle$ or $|\downarrow\rangle$

In matrix form, $\begin{pmatrix}1\\0\end{pmatrix}$ or $\begin{pmatrix}0\\1\end{pmatrix}$

We could have chosen other ways to describe them as well like on and off, up and down, in and out etc. but 1 and 0 are more useful as we can do calculations with them.

Finding the operators

It means we want to know components of the matrix.

Starting with operator of total spin S^2.

We know the eigenvalue of the total spin already. We also know it can act on any of the eigenvectors of spin, but we have chosen z axis vector to be our basis.

$$S^2 = \begin{pmatrix} a & b \\ c & d \end{pmatrix}$$

$$\begin{pmatrix} a & b \\ c & d \end{pmatrix}\begin{pmatrix} 1 \\ 0 \end{pmatrix} = \frac{3}{4}\hbar^2 \begin{pmatrix} 1 \\ 0 \end{pmatrix}$$

Doing matrix multiplication

$a \times 1 + b \times 0 = 1$

This gives a=1.

$c \times 1 + d \times 0 = 0$

This gives c=0.

S^2 acting on the other z component eigenvector will give

$$\begin{pmatrix} a & b \\ c & d \end{pmatrix}\begin{pmatrix} 0 \\ 1 \end{pmatrix} = \frac{3}{4}\hbar^2 \begin{pmatrix} 0 \\ 1 \end{pmatrix}$$

Solving equations again will give b=0 and d=1.

We now have the total spin vector operator.

$$S^2 = \tfrac{3}{4}\hbar^2 \begin{pmatrix} 1 & 0 \\ 0 & 1 \end{pmatrix}$$

Next, finding operator S_z

$$\begin{pmatrix} a & b \\ c & d \end{pmatrix} \begin{pmatrix} 1 \\ 0 \end{pmatrix} = +\tfrac{\hbar}{2} \begin{pmatrix} 1 \\ 0 \end{pmatrix} \text{ and}$$

$$\begin{pmatrix} a & b \\ c & d \end{pmatrix} \begin{pmatrix} 0 \\ 1 \end{pmatrix} = -\tfrac{\hbar}{2} \begin{pmatrix} 0 \\ 1 \end{pmatrix}$$

We can solve the equations in the same way as above giving

$$S_z = \tfrac{1}{2}\hbar \begin{pmatrix} 1 & 0 \\ 0 & -1 \end{pmatrix}$$

Finding operators S_x and S_y is a bit tricky as we do not know the eigenvectors of x and y. We arbitrarily chose the z eigenvector.

We invent new operators that flip the eigenvectors.

$$S_+ = S_x + iS_y$$
$$S_- = S_x - iS_y$$

Why we chose this structure. The honest answer is this is what works. It is a trick that gives the right answer. It the same trick used to solve harmonic oscillator and angular momentum. So, once we find a trick, we use it again and again.

Adding and subtracting the operators gives the solution of the eigenvectors.

$S_x = \frac{S_+ + S_-}{2}$ and $S_y = \frac{S_+ - S_-}{2i}$

If we know the ladder operators, we can find the x and y component vectors.

What do these ladder operators do?

$S_+|\downarrow\rangle = \hbar|\uparrow\rangle$

$S_+|\uparrow\rangle = 0$

$S_-|\uparrow\rangle = \hbar|\downarrow\rangle$

$S_-|\downarrow\rangle = 0$

$S_+|\downarrow\rangle = \hbar|\uparrow\rangle$ means $\begin{pmatrix} a & b \\ c & d \end{pmatrix}\begin{pmatrix} 0 \\ 1 \end{pmatrix} = \hbar\begin{pmatrix} 1 \\ 0 \end{pmatrix}$

Similarly, you have other equations and if you solve them as we did earlier, the matrices are easily calculated.

$S_+ = \hbar\begin{pmatrix} 0 & 1 \\ 0 & 0 \end{pmatrix}$ and $S_- = \hbar\begin{pmatrix} 0 & 0 \\ 1 & 0 \end{pmatrix}$

Once we have the ladder operators, S_x and S_y become

$S_x = \frac{\hbar}{2}\begin{pmatrix} 0 & 1 \\ 1 & 0 \end{pmatrix}$ and $S_y = \frac{\hbar}{2}\begin{pmatrix} 0 & -i \\ i & 0 \end{pmatrix}$

Excluding the factors of $\hbar$, the matrices are usually denoted by the letter sigma σ.

$\sigma_x = \begin{pmatrix} 0 & 1 \\ 1 & 0 \end{pmatrix}$ $\sigma_y = \frac{\hbar}{2}\begin{pmatrix} 0 & -i \\ i & 0 \end{pmatrix}$ $\sigma_z = \begin{pmatrix} 1 & 0 \\ 0 & -1 \end{pmatrix}$

These are called the Pauli spin matrices.

So far, we have found the operators of total spin and x, y and z components. We also have the eigenvector of z component.

What's left?

The eigenvectors of x and y are still to be determined. But once we have the operators, it is not too difficult.

$$S_x \begin{pmatrix} a \\ b \end{pmatrix} = +\frac{\hbar}{2} \begin{pmatrix} a \\ b \end{pmatrix} \text{ or } S_y \begin{pmatrix} b \\ a \end{pmatrix} = -\frac{\hbar}{2} \begin{pmatrix} b \\ a \end{pmatrix}$$

Putting the value of S_x and solving the equations leads to the two eigenvectors of the x component of spin

$$\begin{pmatrix} \frac{1}{\sqrt{2}} \\ \frac{1}{\sqrt{2}} \end{pmatrix} \text{ and } \begin{pmatrix} \frac{1}{\sqrt{2}} \\ \frac{-1}{\sqrt{2}} \end{pmatrix}$$

Similarly, y component eigen vectors are

$$\begin{pmatrix} \frac{1}{\sqrt{2}} \\ \frac{1}{\sqrt{2i}} \end{pmatrix} \text{ and } \begin{pmatrix} \frac{1}{\sqrt{2}} \\ \frac{-1}{\sqrt{2i}} \end{pmatrix}$$

In case you are wondering what's $\frac{1}{\sqrt{2}}$ doing here. It is just a normalization constant. The probability of finding positive and negative eigenvalues has to be 1.

$$\left| \frac{1}{\sqrt{2}} V^+ \right|^2 + \left| \frac{1}{\sqrt{2}} V^- \right|^2 = 1$$

So, probability of getting $+\frac{\hbar}{2}$ and $-\frac{\hbar}{2}$ is ½ and ½ or 50 % each.

What if we know the operator but not the eigenvector or eigenvalues?

$$S_x \begin{pmatrix} x \\ y \end{pmatrix} = \lambda \begin{pmatrix} x \\ y \end{pmatrix}$$

$$\frac{\hbar}{2} \begin{pmatrix} 0 & 1 \\ 1 & 0 \end{pmatrix} \begin{pmatrix} x \\ y \end{pmatrix} = \lambda \begin{pmatrix} x \\ y \end{pmatrix}$$

The formula of finding eigenvalue λ is important and let's review it

$$\det(A - \lambda I) = 0$$

I is the identity operator. It is like multiplying by 1. Let's solve the equation.

$$\frac{\hbar}{2} \begin{pmatrix} 0 & 1 \\ 1 & 0 \end{pmatrix} - \lambda \begin{pmatrix} 1 & 0 \\ 0 & 1 \end{pmatrix}$$

$$\begin{pmatrix} 0 & \frac{\hbar}{2} \\ \frac{\hbar}{2} & 0 \end{pmatrix} - \begin{pmatrix} \lambda & 0 \\ 0 & \lambda \end{pmatrix}$$

$$\det \begin{pmatrix} -\lambda & \frac{\hbar}{2} \\ \frac{\hbar}{2} & \lambda \end{pmatrix} = 0$$

The determinant of 2 × 2 matrix is simply ab-cd

$$-\lambda \times \lambda - \frac{\hbar}{2} \times \frac{\hbar}{2} = 0$$

$$\lambda^2 - \frac{\hbar^2}{4} = 0$$

$\lambda = +\frac{\hbar}{2}, -\frac{\hbar}{2}$ as expected.

To find the eigenvectors, simply put the value of λ in the eigenvalue equation.

$$\frac{\hbar}{2}\begin{pmatrix} 0 & 1 \\ 1 & 0 \end{pmatrix}\begin{pmatrix} x \\ y \end{pmatrix} = +\frac{\hbar}{2}\begin{pmatrix} x \\ y \end{pmatrix}$$

Doing matrix multiplication, we will get x=y

Vector will be $\begin{pmatrix} 1 \\ 1 \end{pmatrix}$

The equation for the negative eigen value will yield x=-y

So, the other vector will be $\begin{pmatrix} 1 \\ -1 \end{pmatrix}$.

Finally, we insert $\frac{1}{\sqrt{2}}$ to normalize the vectors to keep the total probability 1.

In case you are wondering what, does $det\,(A - \lambda I) = 0$ means?

We start with basics, the Schrodinger equation.

$A\Psi = \lambda\Psi$

$(A - \lambda 1)\Psi = 0$

Either $A - \lambda 1 = 0$ or $\Psi = 0$. Obviously wave function cannot be zero.

That makes $(A - \lambda 1)\Psi = 0$

And in the matrix form we get $det\,(A - \lambda I) = 0$

The determinant represents area or volume enclosed by the vector.

Det [ ]=1 means operator vector is rotated as the area remains the same on rotation.

Det [ ]=0 means area or volume enclosed is zero.

This can happen only when vectors are colinear.

For colinear vectors, no area is enclosed, and the determinant is zero.

The operator is unchanged as it acts on the wave function and gives out eigen value which is simply a number. So, it encloses no area with the eigenvalues

A weighing scale gives various values as a result of the measurement. It's determinant with weight measurements is zero as well. Note eigenvalue is a scalar but with identity operator it becomes a vector that encloses zero area with the operator.

We have so far solved the spin in x, y and z direction. What if we chose any arbitrary direction. The answer lies in finding the operator in the nth direction by taking the dot product. It simply means projecting the x, y and z answer onto the arbitrary direction.

$$S_n = (S_x, S_y, S_z).n$$

Putting in the matrix value of the operators, we get

$$S_n = \frac{\hbar}{2}[n_x \begin{pmatrix} 0 & 1 \\ 1 & 0 \end{pmatrix} + n_y \begin{pmatrix} 0 & -i \\ i & 0 \end{pmatrix} + n_z \begin{pmatrix} 1 & 0 \\ 0 & -1 \end{pmatrix}]$$

$$= \frac{\hbar}{2} \begin{pmatrix} n_z & n_x - in_y \\ n_x + in_y & -n_z \end{pmatrix}$$

It is convenient to replace the matrix elements by spherical coordinates.

$$\frac{\hbar}{2} \begin{pmatrix} \cos\theta & \sin\theta e^{-i\phi} \\ \sin\theta e^{i\phi} & -\cos\theta \end{pmatrix}$$

Then we have to solve the determinant equation to find the eigenvalues and eigenvectors.

The eigenvalues will always be $+\frac{\hbar}{2}$ or $-\frac{\hbar}{2}$

Without going into the detailed calculations again, the eigenvectors will be

$$n_+ = \begin{pmatrix} \cos\theta/2 \\ \sin\theta/2 e^{i\phi} \end{pmatrix} \text{ and } n_- = \begin{pmatrix} \sin\theta/2 e^{-i\phi} \\ \cos\theta/2 \end{pmatrix}$$

These vectors can be used to determine any direction of the spin

e.g. for z direction, $\theta = 0$

$$n_+ = \begin{pmatrix} \cos 0 \\ \sin 0 \end{pmatrix} = \begin{pmatrix} 1 \\ 0 \end{pmatrix} \text{ as expected.}$$

Stern-Gerlach Experiment

This was a land mark experiment which led to the confirmatory evidence of quantum spin. The electrons rotate around the atom, so they possess angular momentum. This leads to the development of magnetic field which is deflected in an external magnetic field. In this experiment, we only choose electrons that have no angular momentum that is to say quantum number $l = 0$. You may ask how an electron has no angular momentum, if it rotates around the nucleus. Again, quantum mechanics is tricky. The electron has wave like nature and the probability is spread around the nucleus like a cloud. You can imagine electron waves going back and forth rather that rotating around the nucleus to give zero angular momentum. We rely on mathematics than our imagination!

In this experiment, electrons are beamed through an external magnet. The magnetic field is made uneven by having sharp edges so that we get a torque on the electrons that leads to deflection from their path. What happens when we beam electrons with $l = 0$? These electrons do not possess any angular momentum so there is no magnetic field generated. They should not be influenced by the external magnet and pass straight through. But the beam is deflected and there were only two lines on the detector slide. These two lines are the eigenvalues $+\frac{\hbar}{2}$ and $-\frac{\hbar}{2}$.

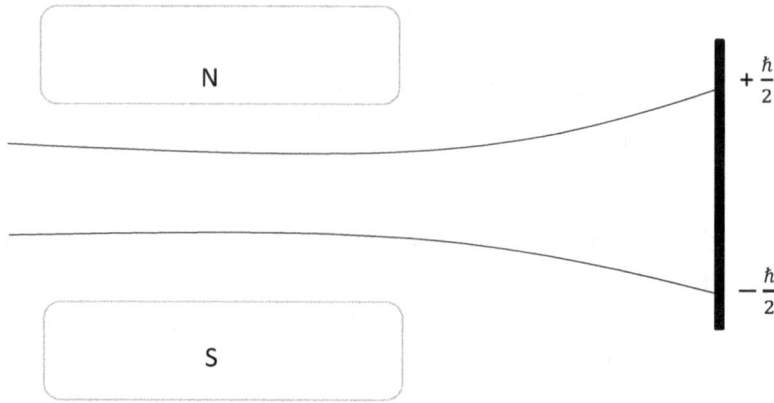

True to the quantum nature of the electrons, there are only two lines. If we had sent classical magnets through the field, we will see continues range of lines from top to bottom. This is because ordinary magnets can be in different orientations like one magnet is at 2 degrees, another one is at 27 degrees and so on. So, each magnet with different orientation gets deflected in its own unique path and fills up the detector plate. But this is not what happens to the electrons. The electron spin is not finalized until a measurement is done. Before the measurement, it is in superposition of all orientations. Only when measurement is made, it picks an orientation. In this case we are doing measurement in the z direction, the electron spin superposition collapses to either $+\frac{\hbar}{2}$ or $-\frac{\hbar}{2}$.

The relationship of spin with its magnetic moment is as follows

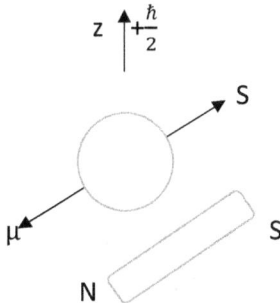

The spin vector is opposite to the magnetic moment in the electron. It points along the south pole of a classical magnet. The magnetic moment is given by formula

$$\mu = -\frac{e}{2m}S$$

The electron experiences torque, that aligns it along the magnetic field so that its South pole is attracted to the North pole of the external magnet.

$$\tau = \mu \times B$$

This torque causes the particle to precess around the magnetic field

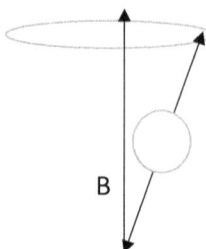

This phenomenon has important practical applications. It is the basis of the MRI machine. Instead of electrons, Hydrogen nuclei are used as they are found in abundance in the human body. Normally the magnetic moments of nuclei are oriented in random directions, canceling each other. But when a strong magnetic field is applied, they start to precess along the magnetic field. The precession frequency is called Larmor frequency. It is given by

$\omega = \gamma B$ where γ is called the gyrometric ratio given by $\mu = -\frac{e}{2m} g$

A pulse is applied in the transverse direction which flattens the precession around the magnetic field.

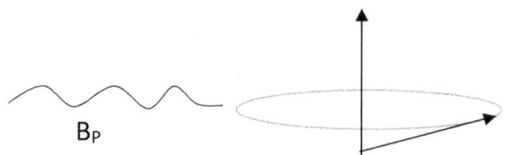

This is a high energy configuration. When the pulse is removed, protons go back to the original precession frequency which is a lower energy state. This process is called relaxation. It causes the emission of radiation. By varying B and pulse, radiation is analyzed. Different parts of the body have varying water and fat in them. Relaxation times are longer for water than fat. This leads to different radiation response which is analyzed to give a useful clinical image. The pulse frequency is the same as the precession frequency. This is called resonance. It causes all the protons to be in phase and pointing in the same direction so that when relaxation happens, the signal is stronger.

Adding spin

We are not done with spin. There are many electrons hanging around and we need to add their spins. In other words, how do spins interact with one another. Like anything in quantum mechanics, things are not straight forward. The quantum weirdness comes even in simple things like adding spin. The reason is that spin is a vector and there is always uncertainty involved in measurements.

Let's add our favored z direction spin components of two electrons. The possibilities are ↑↑, ↑↓, ↓↑, ↓↓.

The problem comes with ↑↓, ↓↑ states. This assumes that we know that electron 1 is spin up and electron two is spin down and vice versa. But unfortunately, quantum mechanics does not allow you to distinguish between electrons. There is no way to tell which is electron 1 or 2. They are identical in every way and if they are close by, their wave functions overlap so their position is not certain as well. More on that later but the point is when two electrons are involved, linear combination of two possibilities have to be made.

How can we make linear combinations?

↑↓ +↓↑ and ↑↓ −↓↑

There is a problem. How do we interpret them?

Do they both mean the same thing?

They do not.

First let's calculate the total z component and total spin of both electrons.

↑↑- Here the combined z component is $\frac{1}{2}+\frac{1}{2}=+1$.

The combined s quantum number is 1. The state is $|s,m> = |1,+1\rangle$

Total Spin is $\sqrt{s(s+1)}\hbar = \sqrt{2}\hbar$

↓↓- Here the combined z component is $-\frac{1}{2}-\frac{1}{2}=-1$.

The combined s quantum number is 1. The state is $|s,m> = |1,-1\rangle$

Total Spin is $\sqrt{s(s+1)}\hbar = \sqrt{2}\hbar$

↑↓ +↓↑- Here the combined z component is $\frac{1}{2}-\frac{1}{2}=0$.

The combined s quantum number is 1. The state is $|s,m> = |1,0\rangle$

Total Spin is $\sqrt{s(s+1)}\hbar = \sqrt{2}\hbar$

↑↓ −↓↑- Here the combined z component is $\frac{1}{2}-\frac{1}{2}=0$.

The combined s quantum number is 0. The state is $|s,m> = |0,0\rangle$

Total Spin is *zero*.

The states ↑↑, ↑↓ +↓↑ and ↓↓ with spin quantum number 1 are called triplet states. They possess the same total spin. The m quantum number ranges from s to -s in integer steps, that is +1,0 and -1.

↑↓ −↓↑ with total spin zero is called the singlet state. Here both s and m quantum numbers are zero.

I am sure you many questions. The first being how on earth ↑↓ +↓↑ can produce total spin number 1 even though z component is 0. How is ↑↓ +↓↑ really different than ↑↓ −↓↑ than arbitrarily giving the results.

It will make more sense if I explain it visually. We are doing vector addition so their direction with respect to one another matters a lot.

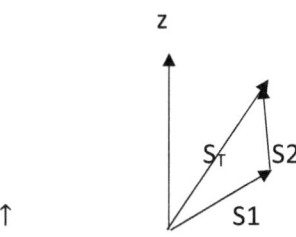

↑↑

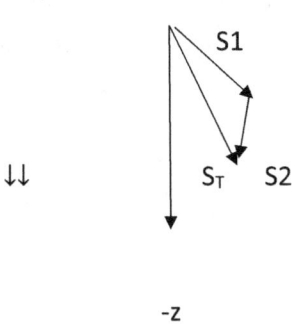

↓↓

−z

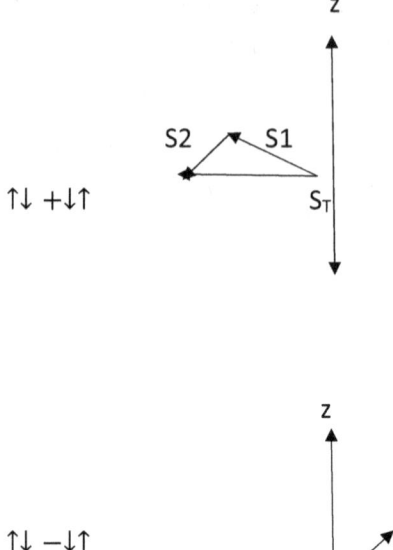

↑↓ +↓↑

↑↓ −↓↑

As you can see the triplet states make the same angle with one another in a way that the spin vectors add onto each other. The ↑↓ +↓↑ state produces zero z axis spin as the vectors do not add in z direction, but they do add in other direction. On the other hand, in the singlet state of ↑↓ −↓↑, the two spin vectors are truly opposite to one another so both z axis and the total spin is zero. This is very different than a classical system where you can add vectors at any angle and get any number of combinations, but quantum spins vectors combine in only fixed discrete ways.

Let me give you a far-fetched analogy that may help to clarify some concepts. Spin can be political too. We will use political spin to explain triplet and singlet states.

↑↑ represents when two republicans align. Things get done for the country(s) as well as for their own base (z direction). We get $|s, m> = |1, +1\rangle$.

↓↓ represents when two democrats align. Things get done for the country(s) as well as for their own base (z direction). We get $|s, m> = |1, -1\rangle$.

↑↓ +↓↑ represents when republicans and democrats align. Things get done for the country(s), but they have to sacrifice their own base (z direction). We get $|s, m> = |1,0\rangle$. These is bipartisan compromise.

↑↓ −↓↑ represents when republicans and democrats anti- align. Things do not get done either for the country(s), or for their own base (z direction). We get $|s, m> = |0,0\rangle$. This is partisan bickering!

The spin structure should fit in nicely with the operators and it does. We need ladder operators to act on these states and they should distinguish triplet from singlet states.

We invent combined ladder operators

$$S_- = S_{-1} + S_{-2}$$

The negative operator flips up spin into down spin. It is made of separate negative operators. The operator of spin 1 only acts on the first electron, leaving the second electron alone and vice versa.

Similarly, a positive ladder operator flips down spin into up spin.

$$S_+ = S_{+1} + S_{+2}$$

S_- acting on ↑↑ will give

$$S_{-1}[↑↑] + S_{-2}[↑↑]$$

S_{-1} will flip the first spin and S_{-2} will flip the second spin

↓↑ +↑↓ wallah! We also have to add $\frac{1}{\sqrt{2}}$ as a normalization constant.

Let the operator act on this state as well.

$S_{-1}[\downarrow\uparrow] + S_{-1}[\uparrow\downarrow] \Rightarrow [0] + [\downarrow\downarrow]$

$+ S_{-2}[\downarrow\uparrow] + S_{-2}[\uparrow\downarrow] \Rightarrow [\downarrow\downarrow] + [0]$

$= \frac{1}{\sqrt{2}}[\downarrow\downarrow] + \frac{1}{\sqrt{2}}[\downarrow\downarrow] = [\downarrow\downarrow]$

S_- acting on $[\uparrow\uparrow]$ m=1

         $[\uparrow\downarrow + \downarrow\uparrow]$ m=0

         $[\downarrow\downarrow]$ m=-1

So, the triplet states fit in nicely into the ladder operators.

What about the singlet state?

S_- acting on $\uparrow\downarrow - \downarrow\uparrow$ will give

$S_{-1}[\uparrow\downarrow] - S_{-1}[\downarrow\uparrow] \Rightarrow [\downarrow\downarrow] - [0]$

$+ S_{-2}[\uparrow\downarrow] - S_{-2}[\downarrow\uparrow] \Rightarrow [0] - [\downarrow\downarrow]$

$[\downarrow\downarrow] - [\downarrow\downarrow] = 0$

You can check that S_+ acting on the singlet state will also give zero. So, the singlet state has s=0 and m=0.

Tensor product notation

To make things more confusing, a short hand notation is used to describe the spin states.

$$\frac{1}{2} \otimes \frac{1}{2} = 1 \oplus 0$$

This notation says that two spin half states can combine to give spin one (triplet state) and spin zero (singlet state).

⊗ is the tensor product. It is used when states are product states which means they can be separated.

$|\uparrow| \times |\uparrow| = |\uparrow\uparrow|$

Two spin up states are product states as they can exist separately.

$\frac{1}{\sqrt{2}}|\uparrow\downarrow - \downarrow\uparrow|$ is not a product state. It is called the entangled state as the components cannot exist independently due to our inability to distinguish electrons. It is called sometimes called a mixed state as our knowledge of its components is incomplete.

Things can get complicated quickly as we add multiple spins. There can be lot of spin combinations to keep track of. There is a formula to write down all possible spin combinations.

The first step is to add the individual spins to form the combined spin.

$j = s_1 + s_2$

Then subtract by 1 till you get to $s_1 - s_2$ where $s_1 > s_2$

These are the possible j's.

Each individual j itself has $(2j + 1)$ states by again subtracting by 1 going from j to -j.

Confusing! Let's do an example, it's not too hard.

$s_1 = \frac{3}{2}, s_2 = \frac{1}{2}$

Let's see how we can combine them.

$j = s_1 + s_2 = \left(\frac{3}{2} + 1\right) - 1$ in integer steps till you get to $\frac{3}{2} - 1$ or $\frac{1}{2}$.

$j = \frac{5}{2}, \frac{3}{2}, \frac{1}{2}$

These are possible three j values.

Each j again has possible m states that go from +j to -j in integer steps

$j = \frac{5}{2}$ has $+\frac{5}{2}, +\frac{3}{2}, +\frac{1}{2}, -\frac{1}{2}, -\frac{3}{2}, -\frac{5}{2}$

Total $(2j + 1) = 6$ states

$j = \frac{3}{2}$ has , $+\frac{3}{2}, +\frac{1}{2}, -\frac{1}{2}, -\frac{3}{2}$

Total $(2j + 1) = 4$ states

$j = \frac{1}{2}$ has , $+\frac{1}{2}, -\frac{1}{2}$

Total $(2j + 1) = 2$ states

In total 12 states are formed.

The same formula can be used to add angular and spin momentum.

$j = l + s$

e.g. $l = \frac{3}{2}, s = \frac{1}{2}$ we will go through the same steps as above.

We also have to add normalization constants to the states. To calculate these constants and form linear combinations is a tedious process. There is a table to quickly read off the constants and form linear combination of these states. It is called the table of Clebsch-Gordon coefficients.

The table is made of sections based on the values of s_1 and s_2.

Let's look at the section of $\frac{3}{2} \times 1$ case we did above.

Here is part of the table

j	$\frac{5}{2}$	$\frac{3}{2}$	$\frac{1}{2}$
m	$+\frac{1}{2}$	$+\frac{1}{2}$	$+\frac{1}{2}$

Coefficients (square root implied)

m_1	m_2			
$+\frac{3}{2}$	-1	$\frac{1}{10}$	$\frac{2}{5}$	$\frac{1}{2}$
$+\frac{1}{2}$	0	$\frac{3}{5}$	$\frac{1}{15}$	$-\frac{1}{3}$
$-\frac{1}{2}$	$+1$	$\frac{3}{10}$	$-\frac{8}{15}$	$\frac{1}{6}$

The table tells us that the combined state $|jm\rangle$ is made up of combination of $|s_1, m_1\rangle |s_2, m_2\rangle$ where s_1 and s_2 are $\frac{3}{2} \times 1$.

$$|\tfrac{3}{2}, +\tfrac{1}{2}\rangle = \sqrt{\tfrac{1}{10}}|\tfrac{3}{2}, +\tfrac{3}{2}\rangle|1, -1\rangle + \sqrt{\tfrac{3}{5}}|\tfrac{3}{2}, +\tfrac{1}{2}\rangle|1, 0\rangle + \sqrt{\tfrac{3}{10}}|\tfrac{3}{2}, -\tfrac{1}{2}\rangle|1, +1\rangle$$

You can look at the table in any standard textbook. It is important to know the underlying concepts. How to read a table is not that important but it does show that calculations can get out of hand as number of states increase.

Density Matrix

We need to learn about the concept of density matrix. It is useful when we mix up spin states. The density matrix ρ is defined as

$\rho = |c_i|^2$ or c^*c where c is the probability coefficient.

What is it good for?

The density matrix can help us calculate the average or expectation value of quantities.

$\langle S \rangle = Tr(\rho S)$

You take the trace of the product of the density matrix and the spin operator whose average eigenvalue you want to calculate.

Taking some trivial examples first. Let's calculate the average value of a dice. The probability coefficient is $\frac{1}{\sqrt{6}}$. It is multiplied by possible dice values (1,2,3,4,5,6) and trace is just adding their products.

$$Tr \begin{pmatrix} \frac{1}{6} & 0 & 0 & 0 & 0 & 0 \\ 0 & \frac{1}{6} & 0 & 0 & 0 & 0 \\ 0 & 0 & \frac{1}{6} & 0 & 0 & 0 \\ 0 & 0 & 0 & \frac{1}{6} & 0 & 0 \\ 0 & 0 & 0 & 0 & \frac{1}{6} & 0 \\ 0 & 0 & 0 & 0 & 0 & \frac{1}{6} \end{pmatrix} \begin{pmatrix} 1 & 0 & 0 & 0 & 0 & 0 \\ 0 & 2 & 0 & 0 & 0 & 0 \\ 0 & 0 & 3 & 0 & 0 & 0 \\ 0 & 0 & 0 & 4 & 0 & 0 \\ 0 & 0 & 0 & 0 & 5 & 0 \\ 0 & 0 & 0 & 0 & 0 & 6 \end{pmatrix}$$

The diagonal elements of the density matrix add up to 1. ($\frac{1}{6} \times 6 = 1$)

Trace of the product is adding the diagonal elements after multiplication

$\frac{1}{6} \times 1 + \frac{1}{6} \times 2 + \frac{1}{6} \times 3 + \frac{1}{6} \times 4 + \frac{1}{6} \times 5 + \frac{1}{6} \times 6 = 3.5$

$\langle Dice \rangle = 3.5$

We will never get 3.5 as a value when we roll out the dice, but it gives us the average of values.

We can play political games with it as well.

If there is a 50% chance of Democrat (+1) or Republican (-1) win then average of winning is

$\langle Election \rangle = Tr(\rho E) = Tr \begin{pmatrix} \frac{1}{2} & 0 \\ 0 & \frac{1}{2} \end{pmatrix} \begin{pmatrix} 1 & 0 \\ 0 & -1 \end{pmatrix} = Tr \begin{pmatrix} \frac{1}{2} & 0 \\ 0 & -\frac{1}{2} \end{pmatrix} = 0$

This just shows that partisanship is a zero-sum game!

There is another way of defining density matrix

$\rho = \Sigma c_i |\Psi_i\rangle\langle\Psi_i|$

The density matrix is the sum of probability coefficients in the direction of the operator.

$|\Psi_i\rangle\langle\Psi_i|$ is the outer product and it projects the results in the operator direction.

It is like saying if car has a speed of 100km/hr., multiplying by $|\Psi_e\rangle\langle\Psi_e|$ will tell us that the direction of 100km/hr is towards the east.

$$\rho = \Sigma c_{dice}|\Psi_{dice}\rangle\langle\Psi_{dice}| = \begin{pmatrix}\frac{1}{\sqrt{6}}\\ \frac{1}{\sqrt{6}}\\ \frac{1}{\sqrt{6}}\end{pmatrix}\begin{pmatrix}\frac{1}{\sqrt{6}} & \frac{1}{\sqrt{6}} & \cdots\end{pmatrix} = 1/6 \text{ matrix we saw earlier.}$$

Let's do some real examples. If we have a pure spin state in the z direction, what is the average spin value? Obviously, we know that it will be 100% as whenever we measure, it will show spin up. But let's do it through density matrix.

Pure spin state $|\uparrow_z\rangle = \begin{pmatrix}1\\0\end{pmatrix}$. The spin eigenvalue is $+\frac{\hbar}{2}$. Let's call it +1.

$$\rho = \Sigma c_i|\uparrow_z\rangle\langle\uparrow_z| = \begin{pmatrix}1\\0\end{pmatrix}\begin{pmatrix}1 & 0\end{pmatrix} = \begin{pmatrix}1 & 0\\0 & 0\end{pmatrix}$$

$$\langle S_z\rangle = Tr(\rho S_z) = \begin{pmatrix}1 & 0\\0 & 0\end{pmatrix}\begin{pmatrix}1 & 0\\0 & -1\end{pmatrix} = \begin{pmatrix}1 & 0\\0 & 0\end{pmatrix} = 1$$

The result is as expected. The result will always show the value of +1 or $+\frac{\hbar}{2}$.

Note for a pure state $\rho = \rho^2$.

Next, we take a mixed state. It means we can have either $+\frac{\hbar}{2}$ or $-\frac{\hbar}{2}$. There is a 50:50 chance to get a spin up or down state.

$$\rho = \frac{1}{2}|\uparrow_z\rangle\langle\uparrow_z| + \frac{1}{2}|\downarrow_z\rangle\langle\downarrow_z| = \frac{1}{2}\begin{pmatrix}1\\0\end{pmatrix}\begin{pmatrix}1 & 0\end{pmatrix} + \frac{1}{2}\begin{pmatrix}0\\1\end{pmatrix}\begin{pmatrix}0 & 1\end{pmatrix} = \frac{1}{2}\begin{pmatrix}1 & 0\\0 & 1\end{pmatrix}$$

$$\langle S_z\rangle = Tr(\rho S_z) = \frac{1}{2}\begin{pmatrix}1 & 0\\0 & 1\end{pmatrix}\begin{pmatrix}1 & 0\\0 & -1\end{pmatrix} = 0$$

Since we have ½ chance of positive and negative values, average is zero.

Note for a mixed state $\rho \neq \rho^2$.

Density matrix is a tool to help calculations for mixed states easier.

Take the case of a complicated mixed state

40% $|\uparrow_z\rangle$ + 35% $|\uparrow_x\rangle$ + 25% $|\uparrow_y\rangle$

$\rho = \rho_1 + \rho_2 + \rho_3$

Any average value can be calculated the same way

$\langle A \rangle = Tr(\rho A)$

The average value can change over time, and is given by

$\frac{d}{dt}\langle A \rangle = \frac{1}{i\hbar}[A, H]$ = Ehrenfest theorem.

Similarly, the density matrix can change over time

$\frac{\partial}{\partial t}\langle \rho \rangle = \frac{1}{i\hbar}[H, \rho]$ = Liouville theorem.

There is sign difference in the commutator between ρ and A. The ρ is usually time dependent but operator A is not.

$\frac{\partial}{\partial t}\langle \rho \rangle = \frac{\partial}{\partial t}|\Psi_i\rangle\langle\Psi_i|$

From Schrödinger equation

$\frac{\partial}{\partial t}|\Psi\rangle = \frac{-i}{\hbar}H|\Psi\rangle$

$\frac{\partial}{\partial t}\langle\Psi| = \frac{i}{\hbar}H\langle\Psi|$

Putting it all together we get $\frac{\partial}{\partial t}\langle \rho \rangle = \frac{1}{i\hbar}[H, \rho]$

There is similar theorem in classical physics $\frac{\partial p}{\partial t} = -\{p, H\} = 0$

Enough with the theorems, what's the big deal?

These theorems show the striking difference between quantum mechanics and classical physics.

In classical physics, the phase space volume does not change. The probability density is constant.

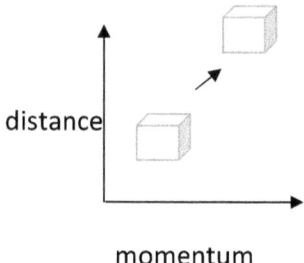

The intuition is simple. If a soccer ball moves through time, it does not shrink on its own. It will go through wear and tear due to friction and heat dissipation. The Liouville theorem does not include effects of friction and heat.

But in quantum mechanics, phase space volume and probability density are not constant.

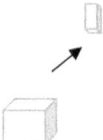

The reason is not difficult to find. The quantum states can exist in superposition where multiple probabilities exist at the same time but due to wave function collapse, only one state remains.

The classical physics is based on set theory, but quantum mechanics is based on vector space or Hilbert's space.

Quantum Entanglement

Entanglement means correlation due to some underlying condition.

Heads and tails of a coin are entangled. If head wins then tail loses.

In sports, winner and loser are entangled. There cannot be a winner without a loser.

$$\Psi_{Toss} = \frac{1}{\sqrt{2}}(\Psi_H + \Psi_T)$$

Ψ_{Toss} is an entangled state, with 50% probability of getting heads or tails.

Let's take a real quantum mechanical example. The conservation of angular momentum requires that if a particle splits into two, then the angular momentum before and after the split should be the same.

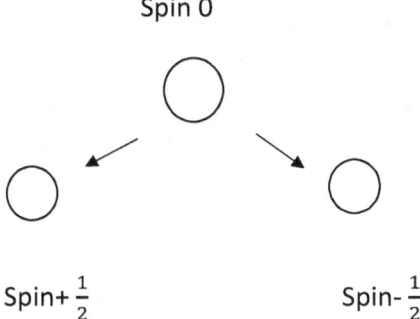

A spin 0 particle will split into two spin half particles. If one particle is spin+$\frac{1}{2}$ then the other particle will be spin+$\frac{1}{2}$. But which of the two will be spin up or down? This sounds like a recipe for entanglement. Whenever there is a choice, a superposition is created. It is our familiar singlet state.

$$\Psi_s = \frac{1}{\sqrt{2}}(\uparrow\downarrow - \downarrow\uparrow)$$

It is a fully entangled state. The constituents cannot be written separately. We have full knowledge of the combined state (spin 0) but are completely uncertain about its constituents (which one is spin up or down).

It is like saying that we are completely certain there will be President chosen in US election but whether it will be Republican, or Democrat is completely uncertain.

I have refrained from philosophizing in this book. But it is too much to resist here.

$$\Psi_{Life} = \frac{1}{\sqrt{2}}(\uparrow_{success}\downarrow_{failure} - \downarrow_{failure}\uparrow_{success})$$

Life is full of ups and downs. You win something, you lose something. You succeed in something, you fail in something. Success and failure are two sides of the same coin and are fully entangled with one another.

Quantum entanglement was made famous by none other than Einstein. He wrote a paper with two other colleagues in 1935. It is known as the famous EPR paper. It is one of the most searched physics papers of all time. The title of the paper was philosophical in nature- Can Quantum-Mechanical description of physical reality be considered complete?

It focused on a paradox known as the EPR paradox. Einstein never liked the random nature of quantum mechanics. Quantum entanglement reached even further, and it seemed that even the basics of special theory of relativity are at risk. This was too much for Einstein to fathom. He had to show the implications of quantum entanglement. It seemed to clash with speed limit of light. If quantum entanglement can break the speed limit of light, then causality is at risk.

We take the example of spin 0 particle again that splits into a singlet state of spin ½ particles. Let's say that the two split particles travel in opposite direction and reach the end of the galaxy.

Across the galaxy

Until the spin of the two particles is measured, both remain in superposition. If particle spin at one end of the galaxy is measured and it shows spin up, the spin of the particle at the other end of the galaxy is instantaneously changed to spin down due to collapse of the wave function. It is called spooky action at distance. There is no regard to the speed limit of light, which is sacred in special theory of relativity. There is a clash of principles here. Einstein postulated that quantum mechanics cannot describe the physical reality. There must be hidden variables that govern the spin of particles that is ignored by quantum mechanics.

The hidden variables mean that spin is predetermined. As the particle splits, the spin of its constituents is determined right away. Quantum mechanics is unable to find the hidden variables, so it uses superposition and collapse of wave function as a substitute to explain results, but it is not describing the physical reality.

In classical physics, when we throw a coin, head and tails are not completely random. If we know hidden variables which means how to throw the coin in a certain way to get heads or tails, in theory it is possible to predetermine the result, the moment you throw the coin. On the other hand, in quantum mechanics, the randomness is paramount and there is no way to get around it.

In 1964, John Bell came up with the idea of Bell's inequality. He proposed that any hidden variable theory must obey Bell's inequality.

What is Bell's inequality?

A (not B) + B (not C) $\geq$ A (not C)

It is based on set theory where states are clearly defined and independent.

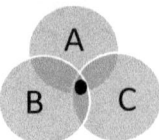

The logic behind it is pretty simple. A not B means A is incomplete. The portion of it is missing (inside B). When you add B not C, the portion of A that was missing is added back except the black area. A not C is more incomplete as the missing part is not added back.

Basically, it is saying more of something ≥ less of something.

If quantum mechanics is based on hidden variables, then it must obey Bell's inequality.

Does quantum mechanics obey Bell's inequality?

Let's calculate it.

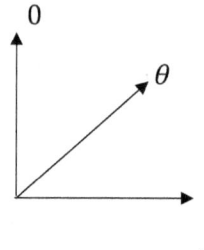

A= Spin up at 0 or z direction.

B= spin up at θ.

Not B means spin down at θ as there are no other allowed values at θ.

C = spin up at 2θ.

Not C = spin down at 2θ.

When we have A Not B on one end of the entangled particle, the other end will have B, and so on

A not B	*Across the galaxy*	*B*
B not C	*Across the galaxy*	*C*
A not C	*Across the galaxy*	*C*

How to calculate the probability of A not B?

$|\langle \Psi | \Psi \rangle|^2 = |\langle A | Not\ B \rangle|^2$

A is spin up eigenvector $|\uparrow\rangle$ in z direction $\begin{pmatrix} 1 \\ 0 \end{pmatrix}$, so taking bra vector $\langle \uparrow | = (1\ 0)$

Not B is n_- vector, we calculated earlier $n_- = \begin{pmatrix} sin\theta/2\ e^{-i\phi} \\ cos\theta/2 \end{pmatrix}$

$|\langle A | Not\ B \rangle|^2 = (1\ \ 0) \begin{pmatrix} sin\theta/2 \\ cos\theta/2 \end{pmatrix} = sin^2\theta/2$

$|\langle B | Not\ C \rangle|^2 = sin^2\theta/2$

The result is similar to A not B as rotating it by θ does not changes the inner product.

$$|\langle A|Not\ C\rangle|^2 = (1\ \ 0)\begin{pmatrix} sin2\theta/2 \\ cos2\theta/2 \end{pmatrix} = sin^2\theta$$

$$sin^2\theta/2 + sin^2\theta/2 \geq sin^2\theta$$

If the angle is small, the small angle approximation $\theta \ll 1$ says

$$sin\theta \sim \theta$$

$$\left(\frac{\theta}{2}\right)^2 + \left(\frac{\theta}{2}\right)^2 \geq \theta^2$$

$$\frac{\theta^2}{4} + \frac{\theta^2}{4} \geq \theta^2$$

$$\frac{\theta^2}{2} \geq \theta^2$$

Obviously, this is not true, so the Bell's inequality is violated.

But why does quantum mechanics violate Bell's inequality?

The reason is that due to quantum entanglement, measurement of one state affects measurement of another state. When we got A not B at one end, this fixed the value of B at the other end. If there is no entanglement, there would be 50% chance of getting spin up and down. But since we have Not B (spin down) at one end, this means spin up at the other end is guaranteed. So, quantum states are not Sets that should obey Bell's inequality. They are vectors in the Hilbert space.

Quantum states are not "set in stone".

Let me give you another political analogy. Suppose we measure the area of countries.

USA (Not Canada) + Canada (Not Europe) ≥ USA (Not Europe)

If measuring the area depends on who measures it e.g. USA may push to extend its boundaries, being a more powerful country, we can see the violation of Bells inequality as well as international treaties!

Does quantum entanglement affect causality?

It is true that wave function collapse is instantaneous and does not care for the speed of light. But thankfully no signal can be transmitted through this process. So, cause and effect relationship are not violated. The person doing measurement on one end will have no idea that B spin was changed in any way by the other end. He will just say that there was a chance to get up value at spin B, so he got it. Only when he does repeat experiments and collects the data and compares with the data from other side, that he realizes that there is correlation between the two particles.

The experiments have been done with photon polarization and there is good evidence that Bell's inequality is indeed violated. It is possible to give an alternate explanation for the findings. The local hidden variables are ruled but not long-distance ones. There is conjecture about the existence of multiverse where all possibilities exist in different universes. I do not want to go into speculative theories. The quantum entanglement is on firm footing and is experimentally proven.

Chapter 7

Quantum Computing

Quantum computing is an exciting concept. It has the potential to fundamentally change computing and revolutionize our lives. The quantum computing can solve difficult computational problems that classical computers struggle with. The field is still in infancy and is long way from becoming a viable alternative to classical computers. Before we go into the basics of quantum computing, we need to do a quick review of how classical computing works. Computer science is a vast field and we will barely be scratching its surface but for our purpose, review of basic concepts will suffice.

The underlying basis of classical computer is a bit. It is the smallest piece of information that serves as the building block of computing. A bit is made up of binary digits, 0 and 1. It contains two pieces of information. The 0 and 1 could be off and on, open and closed, positive or negative etc. You could have more than two states as a basis but 0 and 1 provide a simpler approach to build algorithms, that has become the standard for classical computers.

The bits can be arranged to form logic gates. The basic logic gates are

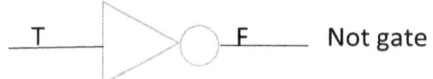 Not gate

It turns True to False and False to True.

```
T  ─┐
     )──  T    And gate
T  ─┘

T  ─┐
     )──  T    OR gate
F  ─┘
```

These gates are basically transistors which open and close switches by manipulating electric current. The gates can be arranged to form a circuit. The algorithms can be written to manipulate circuits. The algorithms form the software and the circuits are the hardware of the classical computer. This is sufficient information for us to move onto quantum computers.

The basic unit of information in quantum computing is quantum bit or qubit. The standard basis of spin state will suffice as a qubit. $\begin{pmatrix}1\\0\end{pmatrix}\begin{pmatrix}0\\1\end{pmatrix}$ represents one qubit where superposition exists whether we will get spin up or down in the z direction. As usual we get spin up and half with certain probability. Qubit has two pieces of information. When we combine two qubits, we get our familiar triplet($\uparrow\uparrow, \uparrow\downarrow +\downarrow\uparrow$, $\downarrow\downarrow$) and singlet ($\uparrow\downarrow -\downarrow\uparrow$) states. Here we have 4 pieces of information. Similarly, 4 qubits have 8 pieces of information, so there is an exponential rise in the information contained in the qubits. The logic gates can be formed from the spin quantum states. The Pauli spin matrices serve as logic gates.

Pauli X gate- like NOT gate

$|0\rangle \rightarrow |1\rangle, |1\rangle \rightarrow |0\rangle$

Here the Pauli x matrix acts on our standard z direction vector and flips it.

$$\begin{pmatrix} 0 & 1 \\ 1 & 0 \end{pmatrix} \begin{pmatrix} 1 \\ 0 \end{pmatrix} = \begin{pmatrix} 0 \\ 1 \end{pmatrix}$$

CNOT-controlled NOT gate

It is a conditional gate where first qubit acts as a control e.g.

If first qubit is 1 then second qubit is acted on by the operator $|10\rangle \rightarrow |11\rangle$

If first bit is 0 then the second qubit is left alone. $|01\rangle \rightarrow |01\rangle$

Hadamard gate

It acts on qubits and convert them into superposition states.

$H|0\rangle \rightarrow \frac{|0\rangle+|1\rangle}{\sqrt{2}}$

$H|1\rangle \rightarrow \frac{|0\rangle-|1\rangle}{\sqrt{2}}$

I do not want to make an exhaustive list but rather give you a flavor of the quantum logic gates.

What's the advantage of quantum computers over classical computers?

This is the heart of the problem. Why bother with quantum computing if there is nothing to be gained.

Classically information contained= $2n$ bits

Quantum information= 2^n bits

For n=100, we are looking at 200 classical bits but 2^{100} quantum bits of information.

But the real problem is how to harness the quantum information. The information is contained in the superposition states, but that information is not accessible. When we do measurement, superposition collapses and we are left with 0 and 1 classical bits only. So, quantum computers are not better than classical computers in everything, only certain set of problems are solved better by quantum computers. If there are infinite paths to go from A to B, then quantum computers have an edge over the classical computers. The classical computer will take one path at a time to figure out the correct answer, which is time consuming. The quantum computer will map out all the paths due to superposition and gives the correct answer by the collapse of the wave function, which is much faster.

The practical problem where quantum computers have proven to be useful is prime number factoring. If N is any number, then it can be written as a product of unique prime numbers.

$N = p \times q$

e.g. $15 = 3 \times 5$

As N gets very large, p and q become very hard to find. This is the basis of encryption. Our passwords and accounts are encrypted using prime factorization where p and q are kept secret. If a hacker can find out the value of p and q, then the encryption can be broken, and accounts hacked. So, N is chosen to be a very large number for security reasons.

The algorithm to find the prime factors of a large odd composite number is done through Shor's algorithm. Before we get into the details of the algorithm, we need to be familiar with modular arithmetic.

The modular arithmetic maps one set of numbers onto another set. It is similar to how clocks work.

12

6

1 2 3 4 5 6 7 8 9 10 11 12 1 2 - Clock set

1 2 3 4 5 6 7 8 9 10 11 12 13 14 -Time set

$13 \equiv 1 (Mod 12)$

The above equation says that 13 is congruent to 1 if Mod is 12. This simply means that in a 12-hour clock, 13 corresponds to 1 as clock only goes to 12 and then recycles. Similarly, 14 corresponds to 2 pm and so on.

There is a quick may to make modular equations

$a \equiv b (Mod N)$

$a - b$ is a multiple of N e.g. 13-1=12, which is a multiple of N=12.

a and b have same remainders when divided by n.

The equation can be written in the following way for prime numbers

$a^r \equiv repeating\ b (Mod N)$

If N and a are prime numbers, then b repeats itself.

0 1 2 3 4 5 6 7 1 2 3   Mod

0 1 2 3 4 5 6 7 8 9 10   a

e.g. $2^1 \equiv 2 (Mod 7)$

$\quad 2^2 \equiv 4 (Mod 7)$

$\quad 2^3 \equiv 1 (Mod 7)$

$\quad 2^4 \equiv 2 (Mod 7)$

$2^5 \equiv 4 (Mod 7)$

$2^6 \equiv 1 (Mod 7)$

As you can see, period here is 3.

What do we do with the period?

If r is odd, then go back and find another number a less than N.

If r is even, then check if $a^{\frac{r}{2}} + 1 \not\equiv 0$ (Mod N)

If yes, then p=gcd ($a^{\frac{r}{2}} - 1$, N)

q=gcd ($a^{\frac{r}{2}} + 1$, N)

e.g. N=15, we pick any random number a less than 15, say 2.

We find period is 4 because $2^4 \equiv 1 (Mod 15)$ repeats itself.

$r = 4, \frac{r}{2} = 2$

The root is even, check.

Then $a^{\frac{r}{2}} + 1 \not\equiv 0$, $2^2 + 1 = 5$ and $5 \not\equiv 0 (Mod 15)$ check.

$a^{\frac{r}{2}} - 1 = 3$ and $a^{\frac{r}{2}} + 1 = 5$

Then p=gcd ($a^{\frac{r}{2}} - 1$, N) = (3,15). gcd of 3 and 15 is 3.

q=gcd ($a^{\frac{r}{2}} - 1$, N) = (5,15). gcd of 5 and 15 is 5.

So, 3 and 5 are the prime factors of 15.

In case you have forgotten, gcd means greatest common divider. You find the highest number that can divide the two given numbers.

It is a tedious process, that's why it is used in encryption. If N is very large, time period r gets to be very big and it can take millions of years for the classical computer to factor N.

How can quantum computing help?

Its main use is to cut the time needed to find the period. The quantum computer can do the same factoring problem in few seconds!

This is achieved as quantum states can be in superposition of all the possible values at the same time and on the collapse of the superposition, only give repeating values. The classical computer on the other hand has to go step by step and solve each entry and equation to find the period which is very time consuming.

Input   Output

$2^1 \equiv 2(Mod\,15)$

$2^2 \equiv 4(Mod\,15)$

$2^3 \equiv 8(Mod\,15)$

$2^4 \equiv 1(Mod\,15)$

$2^5 \equiv 2(Mod\,15)$

$2^6 \equiv 4(Mod\,15)$

$2^7 \equiv 8(Mod\,15)$

$2^8 \equiv 1(Mod\,15)$

Input and output are in superposition. We want it to collapse to repeating output term 1. This is called collision where multiple inputs lead to one output.

Once superposition collapses, only input and output values for 1 are left ($2^4 \equiv 1$, $2^8 \equiv 1$ etc.). We select a certain number of terms from the above sequence(M) such that M is divisible by period r. This requires that M is a large number. For simplicity, I will just make up the numbers. Let's say M is 40. So, it is divisible by period r (4). $\frac{M}{r}$=10.It tells us the number of times an output value repeats itself.

We can see that 4th, 8th and so on entries are left with same output 1. To simplify further, we need to shift these entries so that 4th becomes 0, 8th becomes 4 and so on.

4th 8th 12th 16th

0   4   8   12

The advantage of shifting the entries is that we can see more clearly that period is 4 as entries repeat every 4 units. This process is called Fourier sampling. It is like amplifying the signal through Quantum Fourier Transform. The actual process of shifting basis is quite mathematically involved, and I will spare you the details.

We collect the total output. Say for M=40, we will get 1+1+1.... = 10. The important point is that the total output is a multiple of $\frac{M}{r}$ as 10 is a multiple of 10. We keep collecting samples. Next time using output 2, we may get total output 20. This is also a multiple of $\frac{M}{r}$. Similarly, we may get 40 next time as the total output and so on.

Output samples=10,20,40,80 etc.

We find gcd of the above, which is 10. So, there are 10 entries of each output.

$\frac{M}{r}$ = 10 and M=40. This tells us that in a sample of 40 terms, each value repeats 10 times. The period r is obviously 4. It is like saying, how many sunsets are there in a month(M=30)? We calculate the output and find 30 sunsets.

$\frac{M}{r}$ = 30 , r=$\frac{30}{30}$=1. So, there is one sunset a day.

We cheated a bit here as we already knew the period but in reality, numbers involved in encryption are mind bogglingly large, of order 2^{2000} and period could be of order 2^{500}. Once we find the period, we put the period in the algorithm that we discussed earlier to find the prime factors.

Recapping, quantum computers changed the problem of finding the prime factors to finding the period of a function. The classical computers take forever to find the period, but quantum computers do it much faster by using superposition and collapsing to the desired output.

This is a very simplified way to understand how quantum computing works. The work is only half done as we have to design circuits using quantum logic gates like Hadamard gates to put this algorithm into action.

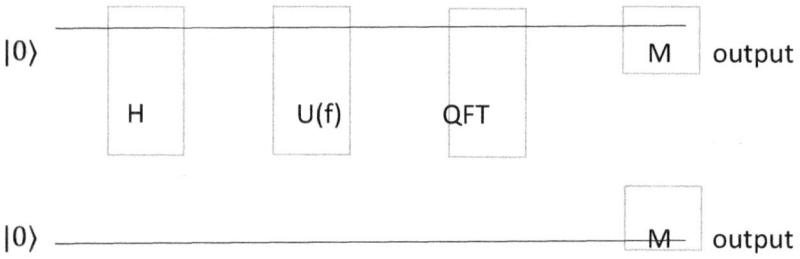

The technical details of the circuit are onerous, and I am not a computer scientist to go into that detail. But roughly, we apply Hadamard gate to the n states so that superposition is established. The U(f) is the mod function we want to calculate. QFT is to change basis for easier calculation of the period and M is the measurement. Hopefully you have some idea about how quantum computing can solve problems that classical computers cannot. There are other algorithms besides Shor's algorithm and quantum computing is a very active area of research.

Why are quantum computers not on our desks?

Notwithstanding the potential of quantum computers, there are many hurdles to overcome before we can see a personal quantum computer. The quantum particles interact with all sort of things, so it's very hard to keep them isolated. The environment interferes with the quantum states, which is called decoherence. The quantum states have to be cooled to near absolute zero temperature and many research labs are designed deep underground to avoid interaction with the environment. At present, quantum computers have only been able to factor numbers like 15,21. They are long way from reaching the numbers involved in encryption. Time will tell if the premise of quantum computing will bear fruit or just frustrate us as it does to physics students.

Quantum Cryptography

If quantum computing has the potential to break any classical password or encryption, quantum mechanics gives us the solution to the problem as well. Quantum cryptography is very secure and impossible to hack secretly. Before we go into the details of quantum encryption, let's review how classical encryption works.

The secret message can be sent between two parties by using cipher text etc. to make it incomprehensible. The way to decode the message is if you have the key. The key helps to decrypt the message. So, sender and receiver need to have keys before sending any secret message and keep the keys secure. If a hacker gets hold of the key, message can be stolen. On a commercial level, it is expensive and time consuming to give secret keys to everyone. A bank has millions of customers, it would be very difficult to give secret key to every customer and expect them to keep it securely. A better method is to have a public key that is accessible to everyone. A customer sends a message through a public key. The bank only has the secret key which it uses to decode the message from the customers. Think of it this way, a bank gives you a suitcase with open lock, you put your secret stuff in the suitcase and lock it. The lock is designed in such a way that no key is needed to lock it. Once the suitcase is locked, the only way it can be opened is through the secret key that bank possesses. This process is called asymmetric key distribution.

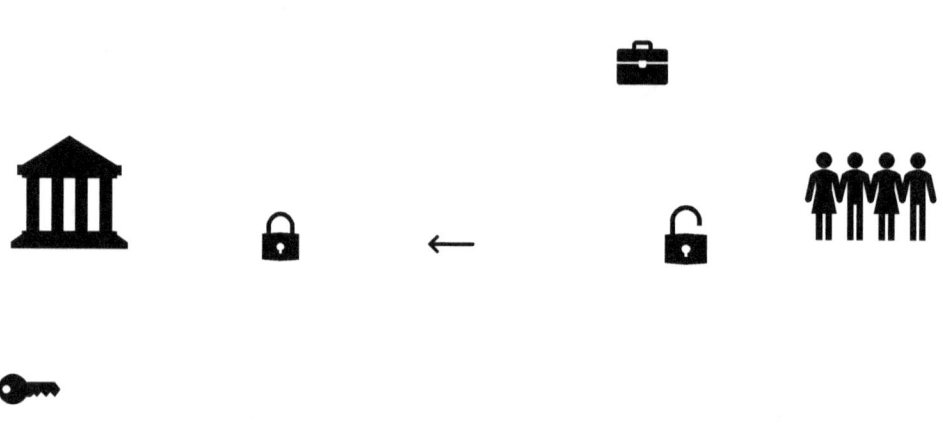

Bank has the key   Public has the lock

RSA is a popular version of public key method. It is used in websites to send data.

It works in the following way.

The public key is a very large number N. It can be 2048 bit or 2^{2048}. N has two unique prime factors p and q. $N = p \times q$

The prime factors of N are the secret key that the bank keeps. The public key N is available to all customers.

The customer wants to send message to the bank, it works as follows

Cipher text=(Message)^e Mod N

Cipher text is unreadable without the key, it is mumbo jumbo of numbers. It is sent to the bank.

Message could be your credit card information etc.

e is the encryption number. It is any random number, there are certain rules for selection which are not that important to know.

What happens at the bank?

The cipher text needs to be decoded. To decode the message, the bank needs a decryption key. It is got by formula

$$e.d \equiv 1[Mod(p-1)(q-1)]$$

Where d is the decryption key.

Decryption key can only be obtained if you know the secret prime factors.

Once you have the decryption key, the message can be decoded

Message$\equiv C^d (Mod\ N)$

The only way for the hacker to get the message is either to steal the key from the bank or factorize N. The time needed to factorize N can be millions of years.

How does quantum cryptography work?

It involves creating a secure channel through which two parties can share the key. The random results of the quantum measurement create the key.

Alice prepares photons through filters so that only select polarization photons come out.

Vertical   Diagonal

When a photon goes through the Vertical-Horizontal filter, it can come out only as

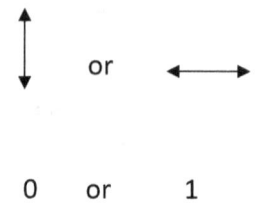

0   or   1

When a photon goes through the diagonal filter, it can come out only as

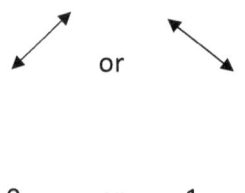

0   or   1

Both states have 50% probability of coming out.

Alice produces a message with 6 photons after passing through the filters.

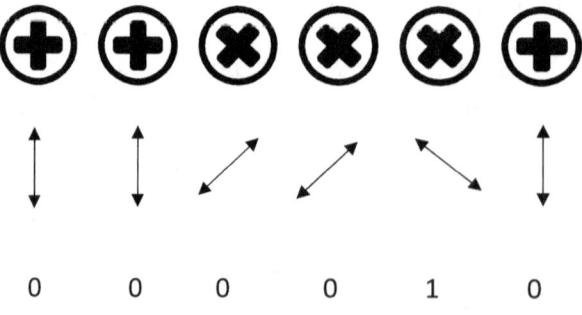

0   0   0   0   1   0

The photons that come out of the filters become the message and are sent to Bob.

Bob has no information about the filters used by Alice, so he applies the filters randomly and gets different results than Alice. So, the message is obviously changed once Bob's filters are applied. If Bob is lucky and applies the same filter as Alice, the photon of the message is unchanged e.g.

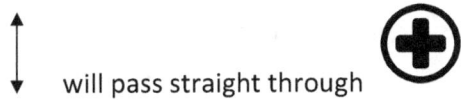 will pass straight through

But if Bob uses a different filter than Alice used to prepare the photons, the message will be changed

will not pass through

The photon will be converted into ↗ or ↖ with equal probability.

The Bob now applies his filters to all six photons sent by Alice

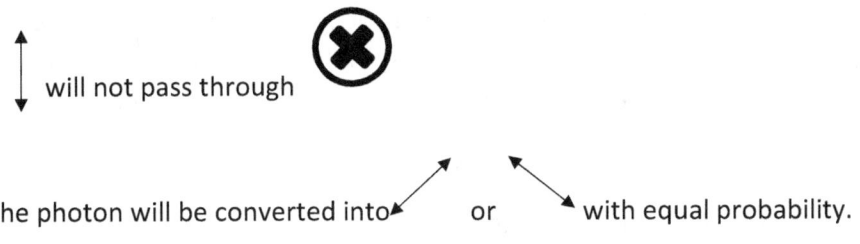

Let's say the following photons come out

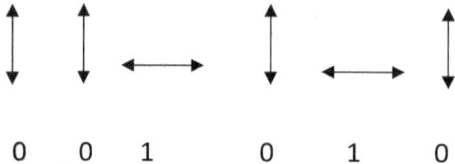

0   0   1   0   1   0

You can see that the vertical photons pass straight through the vertical polarizer. Only diagonal photons are changed into vertical or horizontal randomly with certain probability.

Bob calls Alice on the phone (public channel) and shares his order of filters. Alice and Bob now know which filters Bob got wrong and the results of the wrong filters are discarded. The remaining photons that come out of matched filters become the key. The photons at 3,4,5 positions are discarded as these are the positions where Bob used the wrong filter. The 1,2 and 6th photon are kept as the shared key.

Now both Alice and Bob have the same key (000).

What if a hacker intercepts the message?

The hacker runs into a problem, the quantum state cannot be intercepted without disturbing it. If the hacker uses the same filter as Alice by luck, then the photon will pass through without disturbing it and Bob will not notice if hacker intercepted the message. But luck does not last long as the message contains lots of photons and it is impossible for the hacker to get the exact same polarizer as Alice's every time. If he chooses a wrong polarizer, he will convert a vertical/horizontal photon into a diagonal one. The hacked diagonal photon will go to Bob. The Bob uses the same polarizer as Alice but now since hacked photon is diagonal, the result will match with Alice only half the time.

⊕ ↕ sent by Alice, is intercepted by hacker who uses ⊗ and the photon now becomes ↘ . The diagonal photon now passes through the

Bob's filter ⊕ and he gets ↔ . This is not the same as the original photon sent by Alice. When Bob calls Alice to reconcile the filters, the same filter results should match. But due to hacking they are not matching. To check for hacking, Bob and Alice have to share part of secret key to check for errors like above caused by hacking. Alice says she has 0 (vertical filter) but Bob says he has 1(vertical filter). This is an error, hacking alert! If errors are found, they know that the key is compromised. Then the entire key is discarded. If no errors are found, then the part of the key used for error testing is discarded but the rest of the key is still good to go.

This method of quantum cryptography is called BB84 protocol. It is named after the developers Bennett and Brassard who proposed the protocol in 1984.There are other protocols as well but BB84 is the most popular one.

There are practical problems before quantum cryptography becomes viable. The quantum states are very sensitive to environment, so long distance key sharing is limited. So far, the quantum key has been shared across few hundred km only. There are other sources of error as well. The polarizers may not be exactly vertical or diagonal leading to errors. In theory quantum cryptography has lot of potential but the practical problems need to be sorted out before its ready for prime time.

Chapter 8

Identical Particles

All electrons are alike. There is no way to distinguish one electron from another. God does not discriminate between electrons! This is true for all elementary particles. You may say that even if the electrons are alike, we can keep track of them and still tell them apart based on their location etc. Not so fast, due to uncertainty principle, exact location and momentum is unknown, so tracking is not possible.

Let's say there are two electrons in a box. They are located in east and west directions. After a while, two electrons change their location to north and south locations of the box.

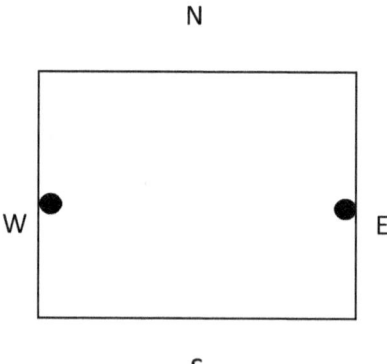

What can we say about the electrons?

If electrons were distinguishable, may be one was red, other was blue, we could say that red electron went to north and blue electron went to south direction of the box. Even if electrons were similar but if we could track them, we would say that west electron went to north and east electron went to south.

But we cannot say that. This is due to uncertainty principle where precise location and direction of the electron cannot be known. We are only allowed to say that two electrons were present in east-west direction and after a while, two electrons were present in north-south direction. Which electron went in what direction is not knowable.

How do we construct a wave function?

$\psi_{(E,W)} = \psi_{1E} \psi_{2W}$

This wave function is wrong for identical particles as it presumes that we can label the electrons. However, it is a perfectly valid function for distinguishable particles, say if one was electron and another was proton.

The correct wave function for identical particles is

$\psi_{(E,W)} = \psi_{1E} \psi_{2W} + \psi_{1W} \psi_{2E}$

It is called symmetric wave function

Similarly, there is another valid function

$\psi_{(E,W)} = \psi_{1E} \psi_{2W} - \psi_{1W} \psi_{2E}$

This is the anti-symmetric wave function.

To normalize, we have to include factor $\frac{1}{\sqrt{2}}$ so that there is equal probability of two combinations.

$\psi_{(E,W)} = \frac{1}{\sqrt{2}} (\psi_{1E} \psi_{2W} - \psi_{1W} \psi_{2E})$

It turns out that plus and minus sign in the wave function is of fundamental importance to distinguish elementary particles.

All elementary particles in the universe fall into two camps

The particles with positive sign in the wave function are called *Bosons*.

The particles with negative sign in the wave function are called *Fermions*.

The sign in the wave function has profound practical implications.

The fermions cannot be in the same state.

$\psi_{(E,W)} = \frac{1}{\sqrt{2}} (\psi_{1E} \psi_{1W} - \psi_{1W} \psi_{1E}) = 0$

You can see that if two fermions are in the same state, wave function is zero. This is the basis of Pauli exclusion principle.

On the other hand, bosons can be in the same state.

Distance between electrons

The separation distance is either W-E or E-W, so we take square of the distance. In QM, average distance is quantity of choice.

$\langle (E - W)^2 \rangle$

$= E^2 + W^2 - 2EW$

Let's look at each individual term.

First of all, our wave function is $\frac{1}{\sqrt{2}} (\psi_{1E} \psi_{2W} - \psi_{1W} \psi_{2E})$

E^2 means what is happening at E alone.

We know intuitively that it means either electron 1 is there or electron two is there.

If we put wave function for E^2, it is again like $(a-b)^2$ or $(\psi_{1E}\psi_{2W} - \psi_{1W}\psi_{2E})^2$

We will have term $(\psi_{1E}\psi_{2W})^2$ and $(\psi_{1W}\psi_{2E})^2$. These terms again say that at E if electron one is there then at W electron 2 will be there and vice versa. The remaining cross terms are zero as there is no overlap in electrons.

Similarly, we will get same terms with W^2.

The interesting term is EW. This is the interacting term.

$(\psi_{1E}\psi_{2W} - \psi_{1W}\psi_{2E})(\psi_{1E}\psi_{2W} - \psi_{1W}\psi_{2E})$

There are square terms again, which are not that interesting.

The cross terms are $\psi_{1E}\psi_{2W}\psi_{1W}\psi_{2E}$ and $\psi_{1W}\psi_{2E}\psi_{1W}\psi_{2E}$.

What to make of the cross terms?

$\psi_{1E}\psi_{2E}$ tells us that electron 1 and electron 2 wave functions are overlapping in the east. The wavefunctions are not probabilities. They can add to cause constructive interference or subtract to cause destructive interference. The wavefunction can be spread out and can be present at two places E and W simultaneously.

Let's call the term $\psi_{overlap}$.

For fermions the overlap term has positive sign because -2EW (- $\psi_{overlap}$)= 2EW + $\psi_{overlap}$

For fermions, the separation distance is

$(E-W)^2 = E^2 + W^2 - 2EW + \psi_{overlap}$

This means distance is increased between fermions as if they are repelling each other due to destructive interference caused by overlap of the functions of both electrons.

For bosons, overlap term is to be subtracted. So, the bosons are closer to each other due to constructive interference.

We have so far included only the space coordinates to label electrons. This is called the space wave function. The spin can be included in the wave function as well. In that case, total wave function needs to be anti-symmetric for the fermions. This means if spins are antisymmetric (with negative sign) as seen in singlet state then the space wave function is allowed to be symmetric as overall function remains anti symmetric.

ψ (antisymmetric)= $(\psi_{1E} \psi_{2W} + \psi_{1W} \psi_{2E})(\uparrow\downarrow - \uparrow\downarrow)$

This is a valid wave function for fermions as symmetric space wave function and anti-symmetric spin wave function leads to the overall function being anti-symmetric under exchange of fermions.

Do you remember Hund's rule for arranging electrons in orbitals?

$\begin{array}{c}\uparrow\ \uparrow\ \uparrow\\ 2p_x\ 2p_y\ 2p_z\end{array}$ is preferred configuration as opposed to $\begin{array}{c}\uparrow\downarrow\ \uparrow\\ 2p_x\ 2p_y\ 2p_z\end{array}$

The reason is that minimum spin or singlet state is anti-symmetric, and this will make space wave function symmetric. The symmetric wave function causes electron distance to be less as we calculated above. This is not desirable as negatively charged electrons do not want to be close to one another.

The higher spin state is symmetric and so space wave function is anti-symmetric and thus increased distance between electrons. The electrons that are not close to one another cause less shielding of the positively charged nucleus. This causes electrons to remain close to nucleus rather than being pushed away by other electrons. This is called screening effect. So, less screening of electrons in the anti-symmetric space wave function is a lower energy state and thus preferred.

How to construct wave function for more than two electrons?

This done by arranging wavefunctions in a matrix and calculating the determinant.

$$\frac{1}{\sqrt{2}}\begin{vmatrix} a & b \\ c & d \end{vmatrix}$$

Determinant ad-cb is our familiar two electron anti-symmetric wave function.

For 3 electrons

$$\frac{1}{\sqrt{3}}\begin{vmatrix} a_1 & b_1 & c_1 \\ a_2 & b_2 & c_2 \\ a_3 & b_3 & c_3 \end{vmatrix}$$

Determinant

$$a_1 \begin{vmatrix} b_2 & c_2 \\ b_3 & c_3 \end{vmatrix} - a_2 \begin{vmatrix} b_1 & c_1 \\ b_3 & c_3 \end{vmatrix} + a_3 \begin{vmatrix} b_1 & c_1 \\ b_2 & c_2 \end{vmatrix}$$

We could extend this process to N number of electrons and it require determinant of $N \times N$ matrix.

You may be wondering that there are gazillions of electrons in the universe. How on earth can we form an anti-symmetric function of all the electrons in the universe?

Thankfully you don't have to. The wave function of an electron is a decaying function with distance. So, the wave function of electron in the lab does not extend much beyond the room. The electron on the moon has no overlap of function with our lab electron. If the electron from our lab moves from one corner to another, we do not have to worry that the moon electron is going to interfere. Thus, we can use the wave function of distinguishable particles to describe the electrons on the earth and the moon.

$\psi_{(E,M)} = \psi_E \psi_M$ is a valid function for non-overlapping electron wave functions.

Solid State Physics

The simple modeling of quantum structure of solids, liquids or gases can reveal intriguing results. The obvious model to use is particle in a box but now we have lots of particles in the box.

The energy of the nth state is related to momentum as follows

$$E_n = \frac{\hbar^2 k^2}{2m}$$

The electrons occupy all kind of energy states. The temperature is nothing but the internal energy of the electrons.

What happens if we cool the electrons down to Absolute Zero temperature?

We should expect that all electrons should go to the minimum energy state or the ground state. But this is not what happens. The electrons cannot occupy same state as per anti-symmetric wave function requirement. So, they occupy one state at a time. This is like going to a theatre, the lowest row seats will be occupied first, followed by higher row seats. Everyone cannot sit in the same seat!

This creates a structure of electrons at Absolute Zero.

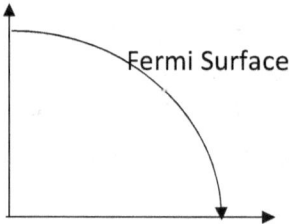

The upper level of the occupied states at Absolute Zero is called Fermi surface. The energy of the uppermost electrons is called the Fermi energy.

The energy states can be represented in terms of momentum, which is called momentum or k space.

$E_f \sim \rho$

The Fermi energy is proportional to density of free electrons. This is obvious, more people packed in a theatre will create more heat.

E_f is unique for different materials. It is analogous to chemical potential. Electrons will flow from material with higher Fermi energy to the material with lower Fermi energy.

Solid State model

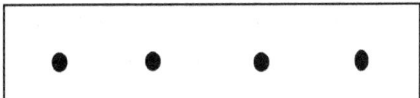

The nuclei are fixed as they are massive as compared to electrons. The electrons move in between nuclei. There is repetitive pattern in the solid. It is as if the entire solid is made up of cells. It is like an apartment building where people move between apartments.

In terms of potential, electrons get attracted to the electric potential of nuclei and potential is same across all nuclei

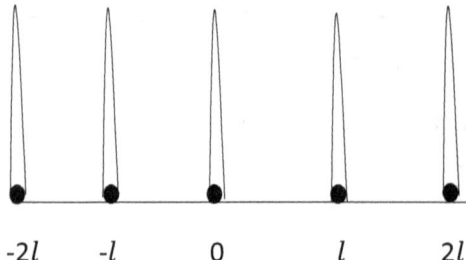

The potential resembles a comb and is appropriately called Dirac Comb potential.

Since potentials are same across nuclei

$V(0) = V(0 + l)$

This is called Bloch Theorem.

The only difference is that that potential carries a phase

$\Psi_{0+l} = e^{ikl} \Psi_0$

This makes things easy, solve the Schrodinger equation in one cell, rest is just a copy of it.

What about the edges?

The last cell at the edge is not connected to anything.

If we take the approximation that there are large number of nuclei in a solid, on the order of Avogadro number, $N \sim 6 \times 10^{23}$ then we just wrap the solid so that edges connect to one another.

$V(0) = V(0 + Nl)$

$\Psi_{0+l} = e^{ikNl} \Psi_0$

For both sides to be equal $e^{ikNl} = 1$

Or $kNl = 0$ or multiple of $2\pi = 2\pi n$

$$k = \frac{2\pi n}{Nl}$$

k is quantized. Recall k is wave vector of electron waves, $k = \frac{2\pi}{\lambda}$. It tells us the direction of the propagation of wave and how are waves wiggling. It is closely related to momentum $p = \hbar k$.

Let's solve the Schrodinger equation

$$-\frac{\hbar^2}{2m}\frac{d^2\psi}{dx^2} = E\psi$$

$$\frac{d^2\psi}{dx^2} = -k^2\psi$$

$$k = \sqrt{\frac{2mE}{\hbar}}$$

We are solving the equation in one cell to start with. Note there is no potential inside the cell.

Solution is

$$\psi = M sinkx + N coskx$$

Next our job is to solve for constants M and N

The solution for the adjoining cell which at distance l is

$$\psi = e^{-iKl}[M sink(x+l) + N cosk(x+l)]$$

At the boundary, x=0 both solutions are equal and ψ goes to zero.

Since cos 0=1 and sin 0=0, sin term vanishes,

$$\psi = 0 + N coskx = 0 + N$$

Equating solutions of both sides

$$N = e^{-iKl}[M sink(0+l) + N cosk(0+l)]$$

Where is potential in all this?

The potential is at the boundary or at $\psi(0)$.

V is zero everywhere except at $\psi(0)$ where it is ∞.

It can be described mathematically as $V(x) = -s\delta$

s is any constant to describe strength of the potential.

This is only an approximation as electron is attracted to nuclei even if it is far away but strength of potential at the nucleus is so large that we can use this simplistic model.

The wave function is continuous at the boundary of cell, electron cannot just disappear.

But $\frac{d\psi}{dx}$ is not defined where V=∞.

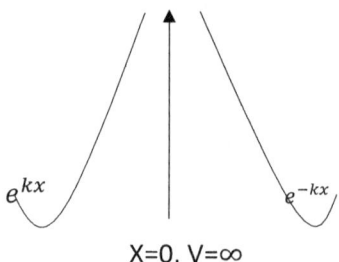

X=0, V=∞

The Schrodinger equation just close to the potential on the either side will be

i.e. from ε to $-\varepsilon$ and taking limit that $\varepsilon = 0$.

$$-\frac{\hbar^2}{2m} \int \frac{d^2\psi}{\partial x^2} + \int V(x) = \int E\psi$$

At $\varepsilon = 0$, $\int E\psi$ term is gone. Remember integral of finite term like ψ is area under curve. If width is zero, there is no area under the curve. The integral is thus zero.

What about $\int V(x)$?

It is equal to s $\psi(0)$ as we defined V this way at x=0.

What is $\int \frac{d^2\psi}{\partial x^2}$?

The second derivative means how first derivative is changing or in other words the difference of derivatives on either side of the potential.

Change of $\frac{d\psi}{dx} = \frac{-2ms}{\hbar^2}\psi(0)$

This is called Dirac Delta function potential.

To get some intuition, think of an apartment building. It has cells or apartments that are similar to one another. There is a gap between balconies. The delta function potential is that gap, where there is danger of falling to the ground.

A person walking in the balcony is well defined. What is not well defined is what happens when he tries to jump across the balcony. In order to solve this, you have to go very close to the edge of the balcony on either side and then evaluate how difficult it is to go across.

Change of $\frac{d\psi}{dx}$ = jump across the balcony.

What will it depend on?

Obviously how deep and wide is the gap. This is the description of the gap potential. That is what delta function potential is trying to describe.

Moving on, we earlier found by equating solutions on adjacent cells that

$$N = e^{-iKl}[M\sin kl + N\cos kl]$$

If we differentiate this equation, it will represent differentiation over the discontinuity at x=0.

Let's equate Dirac delta potential with the above equation

$$\frac{d}{dx}(e^{-iKl}[M\sin kl + N\cos kl]) = \frac{-2ms}{\hbar^2}\psi(0)$$

$\psi(0) = M\sin 0 + N\cos 0 = N$

$$\frac{d}{dx}(e^{-iKl}[M\sin kl + N\cos kl]) = \frac{-2ms}{\hbar^2}N$$

What remains to be done is find $M\sin kl$ in terms of N by basic algebra.

Then there will be N on both sides that will cancel out and thus no constants will be left.

Skipping some algebraic steps, we will be left with equation

$$\cos Kl = \cos kl + \frac{ms}{\hbar^2 k}\sin kl$$

For detailed derivation, see standard textbooks like Griffiths, mentioned in the reference section.

Here K is wave number and small $k = \sqrt{\frac{2mE}{\hbar}}$

The value of $\cos Kl$ is restricted between +1 to -1.

This constraint puts limit on the allowed small k or energy levels.

In discussing Bloch theorem, we found out that wave number can have many values
$$K = \frac{2\pi n}{Nl}$$

$\cos \frac{2\pi n}{N}$ is from +1 to -1.

At multiples of π, there are lot of energy levels that are available to be occupied as N is a large number and we can put any integer value in it.

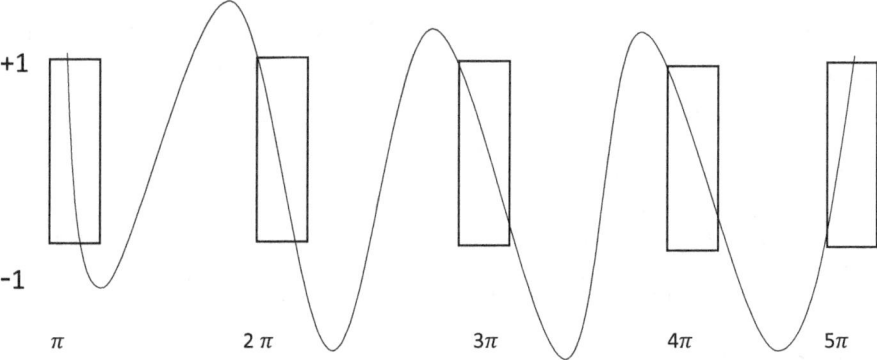

The solutions between +1 and -1 are allowed and form bands, the rest are forbidden or gaps. This is the band theory of solids. The bands that have space to accommodate more electrons allow easy movement across bands are called conductors. The bands that are filled will not allow easy movement across bands. They are called the insulators. The introduction of impurities in the insulator which allows some movement of electrons become semi-conductors.

What to make of the band theory?

The intuition behind it is not difficult to understand. The fundamental equation in band theory is

$$cosKl = coskl + \frac{ms}{\hbar^2 k} sinkl$$

The constraint on $cosKl$ leads to limit on allowed energy levels on right side.

The constraint on $cosKl$ emerged due to folding the edges of solid.

The circle of finite length can accommodate only limited set of waves to fill the circle. The integer waves can fit e.g. n=3,4,5 waves can be made to fit the circle. You cannot fit 3.7 or 7.25 waves exactly into a circle This leads to quantization.

In other words, constraint leads to limited choices.

We see this every day in life. The access to a theatre or stadium is constrained.

There are only certain doors that allow access into the stadium.

What is the response?

People have limited choices to enter, so they form ques in front of doors.

This is band theory in a nutshell.

Similarly, on our highways, speed limits and lanes put limit on vehicles. The conductor highways have plenty of space for vehicles to move in and out of lanes. A bumper to bumper traffic jam is analogous to an insulator.

Quantum Statistical Mechanics

It is the technical name of counting particles. How hard can it be to count particles? This is a question only a naïve person can ask. Statistical mechanics can get very technical and cumbersome. But I will try to keep things simple and recognizable, rather than give you bunch of theorems.

Some basics to start with

Combination

Order is not important. If we have red, blue and green color pencils, just grab them in one go as we don't care in which order we pick color pencils.

Permutation

Order is important. The order in which we can pick three color pencils (R, B, G) will be

RBG BRG GRB

RGB BGR GBR

There are 6 ways to pick 3 objects. In the world of statistics, we call it factorial.

3! = 3×2×1=6

What if we want to pick only 2 color pencils at a time from the 3 color pencils available?

The formula for that is $\frac{N!}{(N-r)!} = \frac{3!}{1!}$ = 6 ways

Leaving R, we can pick BG or GB

Leaving B, we can pick RG or GR

Leaving G, we can pick RB or BR

Total 6 ways again.

What if repetition is not allowed?

Once we pick Blue Green or Green Blue pencils in our hand, what's the difference. Both combinations represent the same state as we can use them similarly to color the picture.

If repetitions are not allowed, then choices will be limited.

Formula is $\frac{N!}{r!(N-r)!} = \frac{3!}{2!1!} = 3$ ways as expected.

Let's make things more complicated

Let's look at three combinations of color pencils

Configuration Q_1 = RRR

Here we come across concept of occupancy number N_n.

R B G

3 0 0

Configuration Q_2 = RRB, BRR, RBR

Occupancy number

R B G

2 1 0

Configuration Q_3 = RBG, BRG, GRB, RGB, BGR, GBR

Occupancy number

R B G

1 1 1

N= total pencils to be selected= 3

d_n = how pencils are used. If we give it to one group (d_1) of people e.g. beginners in coloring, they will use it similarly. If more group of people are involved, degeneracy will be increased accordingly.

Pencils are in a big box. We have to randomly pick three color pencils.

What is the probability to get each configuration?

It will depend on the number of ways each configuration can be chosen as all possibilities are equally likely. So, the configuration that has more ways of being chosen has the highest probability to be picked.

It is obvious that configuration Q_1 can be chosen by one way only, that is if you pick only red colored pencils per attempt.

Q_2 has 3 ways to be chosen and Q_3 has 6 ways to be chosen. So, Q_3 is most likely to be picked of all three configurations due to the fact that there are more ways to choose this configuration as each attempt of picking the pencil is random.

Formula for each configuration Q is

$$N! \prod_{n=1}^{\infty} \frac{dn^{Nn}}{N_n!}$$

I know this looks complicated but not too hard to understand.

N= number of pencils=3

N_n = occupancy

Q_1 the occupancy is $N_1 = 3,0,0$

Q_2 the occupancy is $N_1 = 2,1,0$

Q_3 the occupancy is $N_1 = 1,1,1$

dn^{Nn} = how each configuration is used. If one-way use, then $dn^{Nn} = 1$

e.g.

The possible ways to get RRR is

$N = 3!$

$dn^{Nn} = 1$

$\frac{1}{N_n!} = \frac{1}{3!} \times \frac{1}{0!} \times \frac{1}{0!}$

$Q_1 = 3! \, 1 \left(\frac{1}{3!} \times \frac{1}{0!} \times \frac{1}{0!} \right) = 1$ as expected

The possible ways to get Q_2

$Q_2 = 3! \, 1 \left(\frac{1}{2!} \times \frac{1}{1!} \times \frac{1}{0!} \right) = 3$ ways

The possible ways to get Q_3

$Q_3 = 3! \, 1 \left(\frac{1}{1!} \times \frac{1}{1!} \times \frac{1}{1!} \right) = 6$ ways

The application to quantum mechanics is straight forward. We can choose any number of particles. The energy of a configuration depends on quantum number n. In case of n=2, different values of quantum number l and m represent degenerate states (210, 211, 21-1 and 200) and how many ways each configuration can be achieved will be the most likely configuration.

But there is a problem. The above statistics is only applicable for distinguishable particles. We know that fermions and bosons are identical particles. There is no way of labelling one electron from another. If we have to pick 3 electrons from a box full of electrons, there is no way of arranging them like 123 or 321. Three electrons are three electrons. The only thing to keep in mind is that they cannot be in the same state as antisymmetric wave function forbids it.

Let's say we have three same energy or degenerate states where electrons can be placed. Each state can accommodate one electron only, due to Pauli exclusion principle.

How can we arrange electrons into the states?

For N_1, which means to arrange one electron into three same energy states, choices are

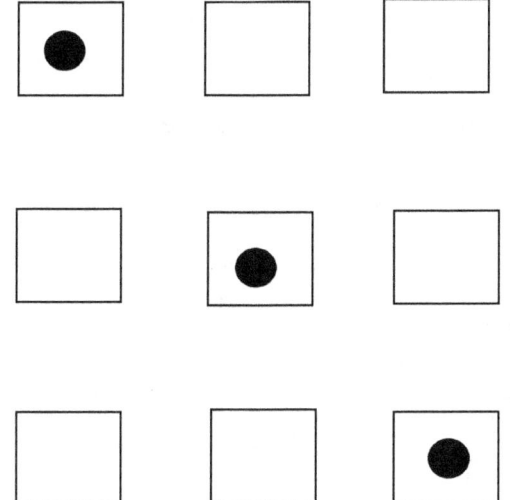

Total choices=3

For N_2, which means to arrange two electrons into three same energy states, choices are

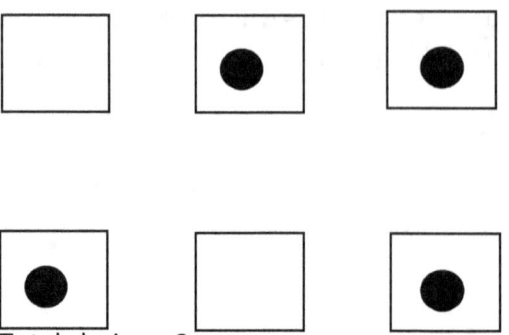

Total choices=3

For N_3, which means to arrange three electrons into three same energy states, choices are

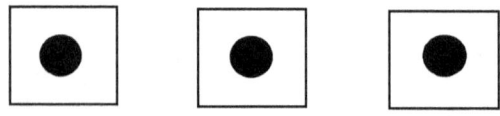

Total choices=1

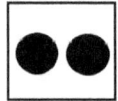

The above state is not allowed for fermions.

For N_4, which means to arrange four electrons into three states.

This is not possible as maximum three electrons can be accommodated by three states.

Let's look at the general formula for Q or configuration of fermions. The total number of electrons that are available to be picked from is not of much importance. So $N!$ is not needed in the formula. What matters is how many electrons (N_n) are picked and distributed into same energy states (d_n).

$$Q = \prod_{n=1}^{\infty} \frac{d_n}{N_n!(d_n-N_n)}$$

$d_n - N_n$ means if electrons are put in energy states, the empty ones need to be counted so that they can be filled.

e.g. for N_1, which means to arrange one electron into d_3 or three same energy states

$$Q = \frac{3!}{1!(3-1)!} = 3$$

Bosonic Statistics

Bosons are friendly. They have no problem in accommodating more than one particle in the same state. This greatly increases our choices to arrange bosons into degenerate states.

For N_2, which means to arrange two bosons into three same energy states, choices are

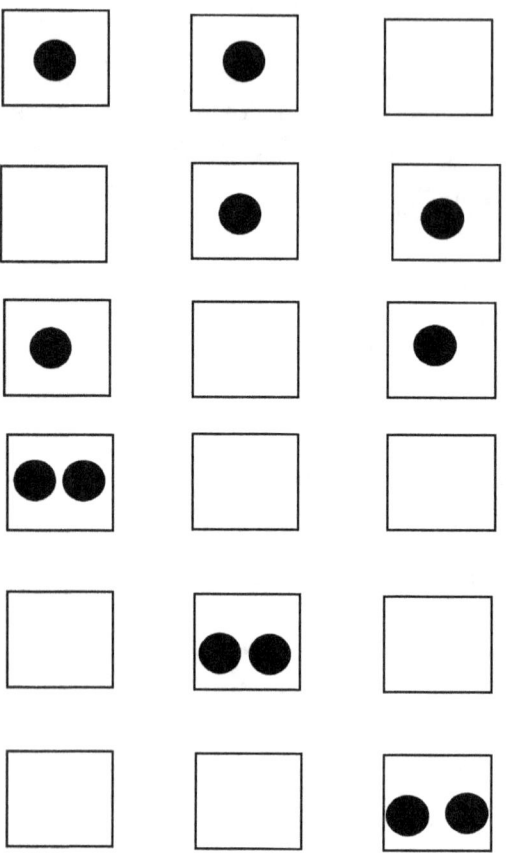

Total choices= 6

This is more than double the choices available for fermions.

How to develop formula for bosons?

Once we put particles in a box, that is fixed. So, the remaining boxes -1 remains.

Add them together. Do the factorial of the end sum. This gives total number of choices. To count boxes, it's convenient to put AND as a marker of separation between boxes.

e.g.

Total choices= Number of particles in box + Number of ANDs -1= 2+2-1=3! Or 6 choices.

What we have done is $(N_n + d_n - 1)!$

This is number of combinations.

Since ordering is not important, we divide by $N_n!\,(d_n - 1)!$

$$Q = \prod_{n=1}^{\infty} \frac{(N_n+d_n-1)!}{N_n!(d_n-1)!}$$

Before I bore you with more statistics, let's look back at the pages of history.

Do you know where does the name boson come from?

It is named after the Indian physicist Satyendra Nath Bose. He first came with the idea that identical particles like bosons should have different statistical rules. If we have two particles that can go into states Heads or Tails, then what are various probabilities?

The probabilities are HH, HT, TH and TT. Each state has ¼ probability of occurrence. But if bosons are identical particles then HT or TH represent same state, so the probability of each state is $\frac{1}{3}$ not ¼.

Let's take a more realistic case. The two bosons can be in ground state or excited state. There is no difference if first boson is in ground state and 2nd boson is in excited state and vice versa. This is because bosons cannot be labeled 1 or 2. The only thing that can be said is that one boson is in excited state and the other one is in ground state. This reduces total number of states from 4 to 3.

Bose wanted to publish his results in prestigious journals. But he faced many hurdles as he was a nobody from a developing country. He took a chance and wrote a letter to Einstein explaining his idea. Einstein recognized the importance of his results. He helped him translate the paper to German and forwarded it for publication. Einstein himself refined the idea in the papers that followed. The statistics of bosons is aptly called Bose-Einstein statistics.

It is noteworthy that Bose never got Nobel Prize for his contribution. It is ironic that Bose-Einstein condensate, a state that was predicted by their statistics was made in the lab in 1990's and several Nobel prizes were awarded to the scientists. I try to see the glass half-full. Even if Bose missed out on the Nobel prize, his name is written as equal with Einstein and all the light (photons are bosons) in the universe is named after him! The fermions on the other hand are name after Enrico fermi and the statistics of fermions is called Dirac-Fermi statistics, in honor of the stalwarts of modern physics, Paul Dirac and Enrico Fermi.

Back to the statistics, we want to find the most probable configuration. To do that, we have to go into the world of Lagrange multipliers.

Some background concepts. A constraint can be put on a function that alters its behavior e.g. you are given one color pencil and asked to fill as many coloring books as you can. One coloring pencil is a constraint and you will have to adjust your coloring style and which books to color first so that you achieve the maximum number of colored pages. The concept of constraint and optimization is important in the field of economics. There are always budget and resources constraints. How maximum profit can be achieved is of great importance.

Let's draw the hypothetical function of a profit curve and budget constraint curve.

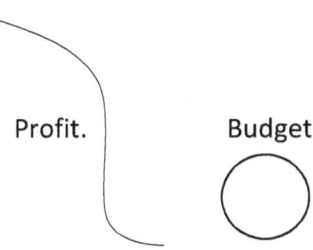

To find the maximum profit, curves have to coincide and be tangential. This would mean at the touching point, solutions to both curves are present, which is what we are looking for

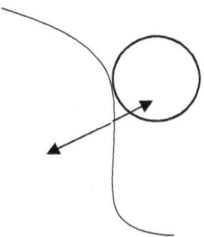

The curve has a gradient(∇)vector which is perpendicular to it. When both curves are tangential, we say the gradient vectors of both curves are aligned and parallel.

$\nabla f \sim \nabla g$

$\nabla f = \lambda \nabla g$

λ is the proportionality constant. It equates the gradient of the function and the constraint. The gradients are parallel, but their magnitude may not be equal.

In economics λ could represent the marginal cost of the constraint. If we relax the constraint or increase the budget, how much can profit increase? This is determined by the value of λ.

$\nabla f - \lambda \nabla g = 0$

There is another way to write it. Let's make up a quantity called L.

$L = f - \lambda g$

$\nabla L = 0$ as $\nabla f - \lambda \nabla g = 0$

The benefit of writing this way is that we can have multiple constraints. In that case, keep on adding constraints.

$L = f - (\lambda_1 g_1 + \lambda_2 g_2 + ...)$

The negative sign in front of the constraint is conventional, you can define it in a different way to get the positive sign.

What are we differentiating with respect to?

The function and constraint can depend on many variables. So, they get differentiated with respect to each variable e.g. x, y and z. This will result in multiple equations which need to be solved to get value of λ and variables.

Let's define function $f = x^3 + y^3$

And constraint $g = x^2 + y^2 = 1$ or $x^2 + y^2 - 1$

To maximize f, $\nabla L = 0$ as $\nabla f - \lambda \nabla g = 0$

L has three variables (x, y and λ) and we need to differentiate with respect to all three variables.

$\nabla f = \frac{\partial x^3}{\partial x} + \frac{\partial y^3}{\partial y} = 3x^2 + 3y^2$. There is no λ dependence in this equation

$\lambda \nabla g = \lambda(\lambda \frac{\partial x^2}{\partial x} + \lambda \frac{\partial y^2}{\partial y} + \frac{\partial \lambda}{\lambda}(x^2 + y^2 - 1)) = \lambda 2x + \lambda 2y + x^2 + y^2 - 1$

Equating equations

$3x^2 + \lambda 2x = 0$

$3y^2 + \lambda 2y = 0$

$x^2 + y^2 - 1 = 0$

Then we have to solve these equations to get the value of variables.

From equation 1, $\lambda = -\frac{3}{2}x$

From equation 2, $\lambda = -\frac{3}{2}y$

Thus $x = y$

From equation 3, $2x^2 - 1 = 0$ or $x = y = \sqrt{\frac{1}{2}}$

Getting back to quantum mechanics, we will use Lagrange multipliers to maximize the configuration of particles.

What are the constraints on a particle configuration?

The total number of particles in all configurations are constant, as they need to be accounted for in the end, no particle can appear or disappear.

Similarly, total energy of particles in all the configurations has to add up to a particular value.

Mathematically it means

$\sum_{n=1}^{\infty} N_n = N$

$\sum_{n=1}^{\infty} N_n E_N = E$

L = Configuration function + Constraints.

$L = \ln(Q) + \lambda_1 (N - \sum_{n=1}^{\infty} N_n) + \lambda_2 (E - \sum_{n=1}^{\infty} N_n E_N)$

We chose to write configuration as log function to make calculus simpler. The log products and divisions become sums and subtractions which are much easier to work with.

The value of Q for distinguished, fermions and bosons need to be plugged in. The rest involves basic differentiation of log functions which I will skip as it is not very illuminating and can be found in standard texts like Griffiths.

We will get the following formulae

$N_n = \dfrac{d_n}{e^{(\lambda_1 + \lambda_2 E_N)} + 1}$ for fermions

$N_n = \dfrac{d_n}{e^{(\lambda_1 + \lambda_2 E_N)} - 1}$ for bosons

$N_1\ E_1 \qquad d_1 \quad d_2 \quad d_3$

$N_1\ E_1 \qquad d_1 \quad d_2 \quad d_3$

N_n means number of particles in a particular energy state. It is the sum of all the particles in the degenerate states available.

What to make of the Lagrange multipliers λ_1 and λ_2 ?

We need to go to the world of thermodynamics. It is a technical and abstract subject. The problem of most probable configuration is well known in thermodynamics. The Q function is replaced by entropy and Lagrange multipliers are called α and β.

α is related to the number of particles in a particular configuration as it represents the constraint $\sum_{n=1}^{\infty} N_n = N$. It is thus related to chemical potential denoted by μ where μ= - αk$_B$T. k$_B$ is the Boltzmann constant and T is the temperature. The particles move from higher chemical potential to the lower potential. It makes sense as higher density of particles will flow to lower density space.

β is related to the coldness of the state. To calculate it, we will have to put value of N_n in both constraints, chose a particular energy state E_N and solve equation.

We will find that $\frac{E}{N} = \frac{3}{2}\beta$.

In classic thermodynamics $\frac{E}{N} = \frac{3}{2}k_BT$

So $\beta = \frac{1}{k_BT}$

β is related to the inverse of temperature. This again is understandable as β represents the constraint $\sum_{n=1}^{\infty} N_n E_N = E$. The internal energy is nothing but the temperature.

In case you forgot, Boltzmann constant is $\frac{R}{N_A}$ where R is gas constant, N$_A$ is Avogadro number.

$PV = nRT$ is the ideal gas law equation.

The Boltzmann constant relates macroscopic pressure and volume to the microscopic energy of particles or temperature. This is just high school physics.

We are more interested in most probable particle state (d$_1$ or d$_2$ etc.) than the state with particular energy (N_n). So, we divide by degeneracy to get the particle state

n= $\frac{N_n}{d_N}$

The formula is a little different as we also include chemical potential and temperature in it.

$$n = \frac{dn}{e^{(\varepsilon-\mu)/k_BT}+1} \quad \text{for fermions}$$

$$n = \frac{dn}{e^{(\varepsilon-\mu)/k_BT}-1} \quad \text{for bosons}$$

What do we learn from these formulae?

They should reveal the characteristics of fermions and bosons that we expect to see from the study of symmetrizing the wave function i.e. fermions do not want to be in the same state, but bosons do.

First take the formula for fermions. Note that the chemical potential at absolute zero temperature is nothing but the fermi energy we studied earlier. The number of particles in each state are higher at lower temperature and lower at higher temperature. This is just common sense, heating up a material will cause particles to spread around and cooling is going to concentrate the material.

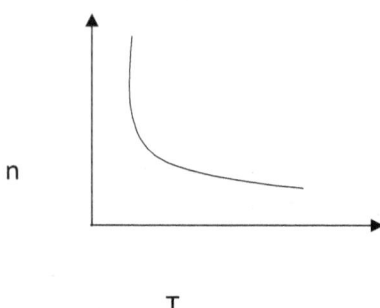

What happens at T=0?

Case 1

$e > \mu$, so $\varepsilon - \mu$ is positive

$$\frac{dn}{e^{(\varepsilon-\mu)/k_BT}+1} \sim \frac{1}{e^{+/0}} \sim \frac{1}{e^{\infty}} \sim \frac{1}{\infty} = 0$$

So, no particle states are occupied with energy more than fermi energy for fermions. This is as expected.

Case 2

$e < \mu$, so $\varepsilon - \mu$ is negative

$$\frac{dn}{e^{(\varepsilon-\mu)/k_BT}+1} \sim \frac{1}{e^{-/0}} \sim \frac{1}{\frac{1}{e^{\infty}}+1} \sim \frac{1}{\frac{1}{\infty}+1} \sim \frac{1}{0+1} = 1$$

So, all particles have energy lower than fermi energy at T=0, bingo!

The fermions do not all collapse into a single state but fill up all the energy states up to the level of fermi energy at absolute zero temperature. This is due to the inability of fermions to be in the same state as per Pauli exclusion principle.

What about bosons?

The bosons have no problem to be in same state. As temperature approaches zero, energy of particles approach zero and chemical potential also approaches zero.

Let's take special case, $\varepsilon - \mu = 0$

$$\frac{dn}{e^{(\varepsilon-\mu)/k_BT}-1} \sim \frac{1}{e^0-1} \sim \frac{1}{1-1} \sim \frac{1}{0} = \infty$$

Particles collapse to a single state. It happens at a critical temperature just approaching zero. It leads to a superfluid with zero viscosity. This is known as Bose-Einstein condensate. Helium was made into a superfluid and it was a significant experimental result, confirming Bose-Einstein statistics.

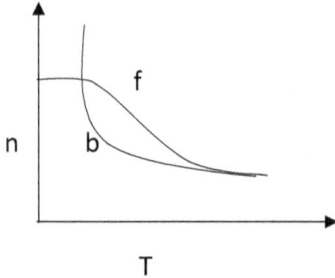

At higher temperature, as density of particles decreases, there is no overlap of wave functions as particles are far apart. This causes bosons and fermions to behave like each other and their curves coincide.

Black body radiation

A black body absorbs all the radiation thrown at it. So, when the black body is heated, it emits radiation of all frequencies. The problem with the emission spectrum is that classically, the intensity of radiation depends on the frequency of radiation. As we heat up the black body, it will emit higher and higher energy or frequency radiation (ultra -violet). This means that the intensity of the radiation will also keep increasing. So, as we heat the black body, we get high energy radiation with increasing intensity. This is a disaster. It is called the ultraviolet catastrophe.

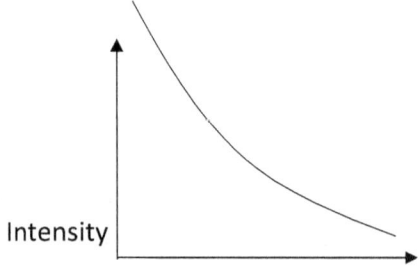

It would be dangerous to sit next to burning coal!

But this is not what is seen experimentally. The intensity does increase with higher frequency but eventually it declines.

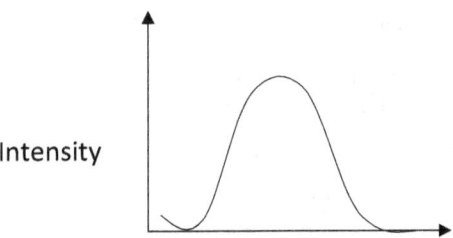

Wavelength (inverse of frequency)

The classical physics was unable to explain black body radiation. This problem led to the invention of quantum mechanics. Max Planck came up with the idea that the radiation can only be emitted in small chunks or quanta. Each quanta of energy are related to the frequency by

$$E = hf$$

He came up with the formula for radiation that was closer to the experimental results, but it lacked sound theoretical basis. But now we know that the chuck of quanta are photons. We can use the Bose-Einstein statistics to explain the black body radiation.

Intensity $\sim nhf \sim \dfrac{1}{e^{hw/k_B T}-1}$

As the temperature increases, the frequency of radiation balances it out ($e^{\frac{w}{T}}$) and ultraviolet catastrophe is avoided.

Entropy

Entropy is an extremely important concept in physics. It is at the core of many fundamental principles. It is usually defined in terms of thermodynamics where it represents all possible configurations. It is a measure of disorder of a system. A macrostate in thermodynamics is one that is measured or observed like volume, pressure etc. The microstate on the other hand is all the possible configurations of a macrostate. A material with a particular temperature(macrostate) can be made of various microstates or possible configurations of the particles. If someone is smiling, that's a macrostate. What makes him, or her smile is a microstate.

$S = k_B \ln W$

The number of ways(W) in which particles can be arranged is entropy(S).

Each state is equally probable.

If we have perfect knowledge of a system, then entropy is zero.

e.g. if a particle can only be in one state then

$S = k_B \ln 1 = 0$

You may wonder why we use logarithm?

The answer is convenience, not due to some deep fundamental principle.

If we have to arrange 10 particles in a box divided into two chambers,

then total number of microstates will be $2^{10} = 1024$.

$\log_2 1024 = 10$.

If we combine similar systems, then adding logs is much more convenient than multiplying 1024×1024.

Sometimes entropy is defined as

$$S = -k_B \sum p_i \ln p_i$$

p_i is the probability of occupying a specific microstate. The only difference is that not all states have equal probability. This happens when a system is not in an equilibrium. When a hot body comes in contact with a cold body, the heat flows from hot to cold body. The cold body states are more likely to be occupied.

The entropy defined this way is called Gibbs entropy.

The entropy of an isolated system increases over time. This is an extremely important principle. An isolated system becomes more chaotic over time.

This is called 2nd law of thermodynamics.

You may ask what negative sign is doing in entropy when it always increases.

Remember this formula is used when not all states are equally probable. In an isolated system, each state has equal probability $\left(\frac{1}{w}\right)$

$$\ln\left(\frac{1}{w}\right) = -\ln W$$

This negative sign cancels the other negative sign in the formula, so the entropy of an isolated system is always positive.

Flow of time

We have all wondered why can't we go back in time? Wouldn't it be wonderful if we can relive our lives or at least good parts of it!

What stops the time from flowing backwards?

There is nothing in Newton's laws of motion, quantum mechanics or Einstein's relativity theory to prevent time from flowing backwards. The answer lies in the 2nd law of thermodynamics. The universe is an isolated system as there is nothing else as far as we know. In an isolated system entropy always increases over time. In an irreversible process, entropy always increases. This means the time in the universe cannot go backwards. Time flowing backwards will decrease entropy, which is not allowed.

Why does entropy always increase in an isolated system like universe?

In other words what is the basis behind the 2nd law of thermodynamics. We have already seen that entropy is the measure of various possibilities. The number of possibilities always increase. If this can happen, then that can also happen, and the game goes on forever. There is nothing to stop the growing of possibilities. This causes entropy to increase and thus fixes the direction of time.

The universe started with a big bang, in a state of low entropy. It stated from a point then expanded to form black holes, galaxies, stars, planets and ultimately life. The universe is getting more complicated with passage of time. The entropy of the universe is increasing, and the arrow of time is one directional. Do not confuse our everyday use of word chaos with entropy. Think of number of possibilities as a measure of entropy. The formation of stars from dust clouds does not decrease entropy as number of possibilities keep growing. Another common mistake is saying that when we clean up our room, don't we decrease the entropy? After all we have arranged things in order. The mistake is we have not considered the work needed. We used our brain, used a vacuum machine to clean the room which taken together increases the entropy.

Chapter 9

Approximation Methods

Things are never simple in quantum mechanics, or for that matter, in real life. There are too many factors influencing each other. It is difficult to track them and get exact results. Many times, you have to work with limited information and do the best you can. The Schrodinger equation is very difficult to solve if we add more variables. This is when perturbation theory comes handy. Perturbation is like disturbance in the smooth potential function. It is typically small compared to the original function for the perturbation theory to work. We will first work with disturbance that does not depend on time or time independent perturbation theory.

As always, start with the eigenvalue equation

$H\Psi = E\Psi$

The Hamiltonian gets disturbed by small perturbation

$H_0 + \lambda V$

λ represents how strong is the perturbation, can go from 0 to 1. It is like a volume knob, you can adjust it to make perturbation noise louder or quieter.

This disturbance will change Ψ and energy values.

$\Psi = \Psi_0 + \lambda \Psi_1 + \lambda^2 \Psi_2 \ldots$

$E = E_0 + \lambda E_1 + \lambda^2 E_2 \ldots$

Higher order terms are needed if we want more accuracy.

Ψ = original function + small change + very small change + very very small change and so on.

Putting back these into the eigenvalue equation, we get

$(H_0 + \lambda V)(\Psi_0 + \lambda \Psi_1 + \lambda^2 \Psi_2) = (E_0 + \lambda E_1 + \lambda^2 E_2)(\Psi_0 + \lambda \Psi_1 + \lambda^2 \Psi_2)$

Rearranging in order of $\lambda's$

$H_0\Psi_0 + \lambda(H_0\Psi_1 + V\Psi_0) + \lambda^2(H_0\Psi_2 + V\Psi_1) = E_0\Psi_0 + \lambda(E_0\Psi_1 + E_1\Psi_0) + \lambda^2(E_0\Psi_2 + E_1\Psi_1 + E_2\Psi_0)$

We have ignored λ^3, λ^4 terms as second order is good enough.

If $\lambda = 0$,

We get back our original equation $H_0\Psi_0 = E_0\Psi_0$

First order λ, equation is

$H_0\Psi_1 + V\Psi_0 = E_0\Psi_1 + E_1\Psi_0$

Second order λ, equation is

$H_0\Psi_2 + V\Psi_1 = E_0\Psi_2 + E_1\Psi_1 + E_2\Psi_0$

We will only be dealing with first order perturbation theory as it's good enough for us.

What do we do next?

The default trick in quantum mechanics is to take the inner product and use normalization condition.

Let's take inner product with Ψ_0 using Dirac notation

$\langle\Psi_0|H_0|\Psi_1\rangle + \langle\Psi_0|V|\Psi_0\rangle = E_0\langle\Psi_0|\Psi_1\rangle + E_1\langle\Psi_0|\Psi_0\rangle$

E_0, E_1 can come out of the inner product as they are just numbers. Note V cannot come out of the inner product as it is an operator and affects inner product.

H_0 can be replaced by E_0 and come out of the inner product

$$E_0 \langle \Psi_0 | \Psi_1 \rangle + \langle \Psi_0 | V | \Psi_0 \rangle = E_0 \langle \Psi_0 | \Psi_1 \rangle + E_1 \langle \Psi_0 | \Psi_0 \rangle$$

First terms on either side cancel, $\langle \Psi_0 | \Psi_0 \rangle = 1$ as per our normalization condition.

We are left with

$$E_1 = \langle \Psi_0 | V | \Psi_0 \rangle$$

This is the first order correction to energy.

Final energy = $E_0 + E_1$

The intuition behind it is straight forward. Let's say we want to know how Apple stock price is determined. Ψ_0 is % of apple stock that is held by big institutions. P_0 is the price on any given day. The potential in this care would be big investors.

H= Big institutional investors like index funds, mutual funds etc.

Small perturbation V = individual or retail investors.

As usual big investors call the shots.

What is the influence of small investors?

$$P_1 = \langle \Psi_0 | V | \Psi_0 \rangle$$

This equation says that price correction is still determined by original function. The stock held by big institutions is what matters. When price changes, it is determined by interaction of Apple stock held by big institutions (Ψ_0) with the mentality of small investors(V). This is because on any given day, most of stock available for trading is dependent on big institutional (Ψ_0) activitiy.

What if you want to be very accurate?

Day traders may want to know every cent fluctuation in the stock price.

Then second order price changes need to be determined.

$P_2 = \langle \Psi_0 | V | \Psi_1 \rangle$

In this case, we have to include changes to apple stock caused by small investors ($|\Psi_1\rangle$).

I hope this analogy makes sense to you, but do not take it literally.

Moving on, what's left to do?

First order correction to Ψ or Ψ_1.

The calculation is little tedious. We have to introduce more superscripts to the equations. This is just to label things.

We label our original state function as Ψ_0^n. When taking inner product, we will use a different function Ψ_0^m. You can think of n as ground state function and m as first excited state. The point is they represent different states.

Let's write inner product with new superscripts.

$E_0^m \langle \Psi_0^m | \Psi_1^n \rangle + \langle \Psi_0^m | V | \Psi_0^n \rangle = E_0^n \langle \Psi_0^m | \Psi_1^n \rangle + E_1^n \langle \Psi_0^m | \Psi_0^n \rangle$

$\langle \Psi_0^m | \Psi_0^n \rangle = 0$ as there is no overlap between these two different states.

$\langle \Psi_0^m | V | \Psi_0^n \rangle = (E_0^n - E_0^m) \langle \Psi_0^m | \Psi_1^n \rangle$

Or

$\langle \Psi_0^m | \Psi_1^n \rangle = \dfrac{\langle \Psi_0^m | V | \Psi_0^n \rangle}{(E_0^n - E_0^m)}$

If we multiply by the identity operator, nothing happens to the function.

$|\Psi_1^n\rangle = \hat{I} |\Psi_1^n\rangle$

It is like multiplying by 1.

Outer product $|e\rangle\langle e|$ is like an identity operator. It is a projection operator and does not affect the function. It is like saying that car is moving in north east direction with a certain speed. If we want to know, how much speed is there in the east direction, we take inner product $\langle e|ne\rangle$ which gives a number say 55km/hr. Then $55|e\rangle$ represents that car is moving 55km/hr in the east direction.

$|\Psi_1^n\rangle = \sum |\Psi_0^m\rangle \langle \Psi_0^m | \Psi_1^n\rangle$

Replacing $\langle \Psi_0^m|\Psi_1^n\rangle$ by the expression we calculated earlier,

$|\Psi_1^n\rangle = \sum |\Psi_0^m\rangle \frac{\langle \Psi_0^m|V|\Psi_0^n\rangle}{(E_0^n - E_0^m)}$

This is the first order correction to the wave function.

Summation $\sum$ is over all the different states. It is like saying we want projection in south, west, and north directions. So, we have to sum over all directions to find the projected car speed.

Note that different states should not be sharing same energy values. If a function represents ground state, another state has to represent excited state.

This is why it is called *non-degenerate perturbation theory*.

If eigenfunctions share the same energies, then $E_0^n - E_0^m = 0$ and the equation will not work.

Think of different functions collectively make a skyscraper. Energy levels represent different floors in the building. Perturbation could a car that accidently hit the building. If the $E_0^n - E_0^m$ is huge then it a big building and correction to the original function is going to be small. It means a car hitting a big skyscraper won't do much damage. But if the energy difference is small then damage could be significant, and first order correction will be more.

$E_0^n - E_0^m = 0$ is like saying that floors on the building are not numbered, then it is harder for you to know size of the building and the damage caused by the perturbation.

$\langle \Psi_0^m | V | \Psi_0^n \rangle$ is the strength of the perturbation. Bigger the perturbation, bigger the correction. A truck hitting the building will do more damage than a scooter.

Degenerate Perturbation Theory

The energy levels are like labels. They help us track different states. If states share same energy levels, it is difficult to distinguish states. It is like saying that we can tell which players belong to a particular soccer team based on the color of their uniform. If all players wear the same uniform, it is going to be much harder to distinguish players.

We begin with two functions sharing the same energy value. Ψ_0^m and Ψ_0^n represent same energy value of E_0.

Since we do not know which of these two states are in the original function as it's hard to distinguish between them, so we have to include them both in the original function.

$$\Psi_0 = \alpha \Psi_0^m + \beta \Psi_0^n$$

α and β represent what percentage of each state is included in the function.

The first order energy correction as discussed before was

$H_0 \Psi_1 + V \Psi_0 = E_0 \Psi_1 + E_1 \Psi_0$

Taking inner product with $\langle \Psi_0^m |$

$\langle \Psi_0^m | H_0 | \Psi_1 \rangle + \langle \Psi_0^m | V | \Psi_0 \rangle = E_0 \langle \Psi_0^m | \Psi_1 \rangle + E_1 \langle \Psi_0^m | \Psi_0 \rangle$

$E_0 \langle \Psi_0^m | \Psi_1 \rangle + \langle \Psi_0^m | V | \Psi_0 \rangle = E_0 \langle \Psi_0^m | \Psi_1 \rangle + E_1 \langle \Psi_0^m | \Psi_0 \rangle$

After cancelling terms

$\langle \Psi_0^m | V | \Psi_0 \rangle = E_1 \langle \Psi_0^m | \Psi_0 \rangle$

Here we need to replace $\Psi_0 = \alpha \Psi_0^m + \beta \Psi_0^n$

$\alpha \langle \Psi_0^m | V | \Psi_0^m \rangle + \beta \langle \Psi_0^m | V | \Psi_0^n \rangle = E_1 \alpha \langle \Psi_0^m | \Psi_0^m \rangle + E_1 \beta \langle \Psi_0^m | \Psi_0^n \rangle$

$\langle \Psi_0^m | \Psi_0^n \rangle = 0$ due to orthogonality which means different states have no overlap, at least in their function.

$\langle \Psi_0^m | \Psi_0^m \rangle = 1$ due to normalization

So, we are left with

$\alpha \langle \Psi_0^m | V | \Psi_0^m \rangle + \beta \langle \Psi_0^m | V | \Psi_0^n \rangle = E_1 \alpha$

Similarly, if we had taken inner product with $\langle \Psi_0^n |$

$\alpha \langle \Psi_0^n | V | \Psi_0^m \rangle + \beta \langle \Psi_0^n | V | \Psi_0^n \rangle = E_1 \beta$

There is a neat way to write these equations

$\alpha V_{mm} + \beta V_{mn} = E_1 \alpha$

$\alpha V_{nm} + \beta V_{nn} = E_1 \beta$

They can be arranged in a matrix form

$$\begin{pmatrix} V_{mm} & V_{mn} \\ V_{nm} & V_{nn} \end{pmatrix} \begin{pmatrix} \alpha \\ \beta \end{pmatrix} = E_1 \begin{pmatrix} \alpha \\ \beta \end{pmatrix}$$

What's the point?

Does it remind you of the eigenvalue problem?

$$V\Psi = E\Psi$$
$$(V - E \times 1)\Psi = 0$$

In matrix form $[\hat{V} - \lambda I] = 0$

E is replaced by λ which is eigenvalue of this matrix. I is the identity operator, it is like multiplying by 1. α, β are eigenvectors.

If you are rusty on details, go back and learn about the eigenvalue problem in earlier chapters.

The matrix and eigenvectors can be extended depending on the degenerate states. $n \times n$ matrix and n vectors can be included e.g. if three states share same energy then there will be 3 × 3 matrix and three state column vector.

e.g. if V matrix is of the form $\begin{pmatrix} a & b \\ c & d \end{pmatrix}$

Then doing matrix multiplication will result in

$$\begin{pmatrix} a - \lambda & b \\ c & d - \lambda \end{pmatrix} = 0$$

Det[$(a - \lambda)(d - \lambda)$ - bc = 0

This is the characteristic equation and the polynomial can be solved to give two roots of λ. Say we got two roots as 1 and 0.5.

What are they good for?

The first order energy correction is

$E_1 = \langle \Psi_0 | V | \Psi_0 \rangle = \lambda V$

Since λ has two roots or values, E_1 will be split into two values.

$E = E_0 + E_1$

So degenerate states do not have same energy but are split into two, due to energy correction.

Let's get some intuition as it feels too mathy!

In our building analogy, all floors have the same number or there is no number written to distinguish floors. So, we need to use different label to distinguish floors. If a car hits the building and causes damage, then only way to tell which floor got more damage is to go from one floor to another and compare. So, if one floor has more damage, another has minimal damage, we can tell that the damaged floor is likely closer to the car impact. This comparison by going from floor to floor is what makes the matrix. The floor numbers are not "good" states to distinguish. The good states are the one with windows where damage can easily be spotted, and they become new eigenvectors. The damage which is different from one window to another, is the new eigenvalue. It causes splitting of energy values which is useful to distinguish one floor from another.

Practical Example

We know that energy for Hydrogen $\sim n^2$

The n quantum number determines energy. Take the case of n=2.

According to quantum number rules, if n=2, l=0 and 1, m=+1,0 and -1.

There are 4 states representing same energy-200(2s state) and 211,210 and 21-1 states(2p).

This means we need 4 × 4 matrix and 4 eigenvectors for the equation. It turns out that due to selection rules, most of the matrix elements are zero except 2.

To see that we study electrical potential as the perturbation

$V = eEz$

To see how selection rules work, we use special tricks

We start with commutation relation

$[L_z, z] = 0$

If we take its expectation value, it should be zero. So, let's take it

$\langle l'm' | [L_z, z] | lm \rangle = 0$

$[L_z, z] = L_z z - z L_z = (m'\hbar)z - z(m\hbar)$, putting it back into equation

$\langle l'm' | [(m'\hbar)z - z(m\hbar)] | lm \rangle$

Or $(m' - m)\hbar \langle l'm' | z | lm \rangle = 0$

This equation says that matrix element is non-zero only if $m = m'$

This leaves only two $\langle 210 | z | 200 \rangle$ and $\langle 200 | z | 210 \rangle$ elements in the matrix as non-zero.

$$\begin{pmatrix} 0 & \langle 200| z | 210 \rangle & 0 & 0 \\ \langle 210| z | 200 \rangle & 0 & 0 & 0 \\ 0 & 0 & 0 & 0 \\ 0 & 0 & 0 & 0 \end{pmatrix}$$

We will need to calculate two non-zero elements by putting eigenfunction from Hydrogen solution and carry out integral. Then solve characteristic equation to get first order energy correction.

The first order correction is $E_1 = \pm 3 e a_0 E$

Here E is the electric field coming from the perturbation potential.

This breaks the degeneracy of two states and this effect is known as Stark Effect.

Trick to use non-degenerate theory

Degenerate theory is cumbersome as we have to do lots of calculations to find all the elements of the matrix and then find eigenvalues. Non-degenerate theory is relatively simple.

If we can find an operator that commutes with perturbation

$[C, V] = 0$

The operator C produces different eigenvalues when it acts on the original states.

$C = g \, \Psi_0^m$

$C = r \, \Psi_0^n$

Where $g \neq r$

Let's allow the states to act on the commutator, which will still be zero meaning both C and V do not affect each other's measurement.

$\langle \Psi_0^m [C, V] \Psi_0^n \rangle = 0$

$[C, V] = CV - VC = 0$

$\langle \Psi_0^m [C\ V] \Psi_0^n \rangle - \langle \Psi_0^m [V\ C] \Psi_0^n \rangle = 0$

Putting in respective eigen values of C, that come out of the inner product

$(g - r)\langle \Psi_0^m |V| \Psi_0^n \rangle = 0$

Since $g \neq r$

$\langle \Psi_0^m |V| \Psi_0^n \rangle = 0$

Going back to degenerate theory equation

$\alpha \langle \Psi_0^m |V| \Psi_0^m \rangle + \beta \langle \Psi_0^m |V| \Psi_0^n \rangle = E_1 \alpha$

So, $\beta \langle \Psi_0^m |V| \Psi_0^n \rangle = 0$

The only term left is

$\langle \Psi_0^m |V| \Psi_0^m \rangle = E_1$

Which is nothing but non-degenerate theory.

The intuition is again finding a new label to distinguish things. If the floors of our building are of different colors then it does not matter that floors are not numbered, we can use color to distinguish floors and can easily tell the damage caused by the perturbation. The C is color operator, it tells color of the floor-red or green as eigenvalues.

$[C, V] = 0$ means that perturbation or the car damage is not affecting the colors of the floors.

$\langle \Psi_0^m |V| \Psi_0^n \rangle = 0$

Just reinforces the fact that car damage is not doing to change the color of these two states.

I hope the analogy gives you some perspective to the logic behind perturbation theory as it is easy to get lost in details of mathematics.

Fine structure of Hydrogen

The energy splitting is based on solving the Schrodinger equation. The electromagnetic potential is used to solve the equation. It divides Hydrogen atom in orbitals with different energies based on quantum number n.

Fine structure of hydrogen explores the finer details of the spectrum. It means adding more variables or perturbations to the potential and solving the energy corrections. This will show that energy levels depend on the factors besides the n quantum number. The perturbation theory will come handy in exploring the fine structure.

Relativistic correction

The basic equation of special relativity is

$$E = \sqrt{p^2c^2 + m^2c^4}$$

We will derive and explore it in later chapters.

For now, this equation can be expanded in Taylor series

$$mc^2 + \frac{p^2}{2m} - \frac{1}{8}\frac{p^4}{m^3c^2}...$$

Here perturbation to the KE $=\frac{p^2}{2m}$ is $\frac{1}{8}\frac{p^4}{m^3c^2}$.

$V = \frac{1}{8}\frac{p^4}{m^3c^2}$

Fortunately, we can use non-degenerate perturbation theory. This is because V commutes with L^2 and L_z and so nlm are good quantum numbers to use.

$[V, L^2] = 0$

$\langle \Psi_0^m [V, L^2] \Psi_0^n \rangle = 0$

$(l^m(l^m+1) - l^n(l^n+1))\langle \Psi_0^m | V | \Psi_0^n \rangle = 0$

Since $l^m \neq l^n$, the other term is zero.

Similarly, $[V, L_z] = 0$

Let's use the non-degenerate theory

$\langle \Psi_0 | V | \Psi_0 \rangle = E_1$

$\langle \Psi_0 | p^4 | \Psi_0 \rangle = E_1$

$\langle p^2 \Psi_0 | p^2 \Psi_0 \rangle = E_1$

From Schrodinger equation

$p^2 \Psi_0 = 2m(E_0 - V_0)\Psi_0$

$\langle E_0 - V_0 \rangle^2 \approx E_1$

The next step is to put the electric potential $V_0 = -\frac{e^2}{4\pi\varepsilon_0}\frac{1}{r}$ into the equation and solve the integral. The result will be

$E_1 \approx E_0^2 \, \alpha^2$ (factors of n and l)

$\alpha = \dfrac{e^2}{4\pi\varepsilon_0} \dfrac{1}{\hbar c}$ is called the find structure constant with value of $\dfrac{1}{137}$

The relativistic correction to the energy is very small, of the order 2×10^{-5}.

The main point is that relativistic correction tells that energy depends on quantum number l as well.

Spin-orbital coupling

The electron is spinning around itself and orbiting the proton. The spin angular momentum and orbital angular momentum interact with each other leading to coupling. It is a perturbation that needs to be solved.

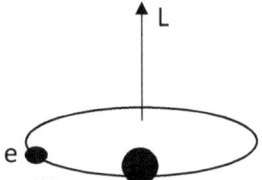

The potential energy generated by the charged spinning electron going around the proton is given by $V = -\mu \cdot B$

B is the magnetic field. It is the field experienced by the electron. It is generated by the proton. From electron point of view, proton is rotating around it. This is like ancient people who thought sun moved around the earth. The moving proton is like a loop of current which generates the magnetic field.

It is given by formula

$$B = \frac{1}{4\pi\varepsilon_0} \frac{e}{mc^2 m^3} L$$

μ is the magnetic moment of the electron. It is caused by spinning of the charged electron. The electron becomes a little magnet. The magnetic moment tells us the orientation and strength of the magnetic field. Its classical expression is $\mu = IA$.

The current in the loop and its area determines the magnetic moment. In classical physics, the magnetic moment generates torque when placed in an external magnetic field as the field tries to align the moment with it. The following expression can be obtained after some elementary calculations.

$$\mu = -\frac{e}{m} S$$

The magnetic moment points in the North pole direction. The magnetic field B is conventionally from North to the South pole. The magnetic moment formula also tells us that electron magnetic moment and spin vectors point in opposite directions.

So, the magnetic field B will try to make magnetic moment point in the same direction as the field.

High energy state   Low energy state

Since B depends on angular momentum L, if L and S are aligned in the same direction then it is a high energy state. If L and S are anti-aligned, then it is a low energy state.

How to add L and S?

L + S is nothing but the total angular momentum J.

J is given by formula

$$J = \sqrt{j(j+1)}\hbar$$

j quantum number is obtained by adding l and s quantum numbers in integer steps

$(l-s) +1 …… (l +s) -1$

Take case of $l = 2$, $s = \frac{1}{2}$

$l + s = \frac{5}{2}$ and $l - s = \frac{3}{2}$

$l + s$ is aligned state, there are total 6 aligned states from $+\frac{5}{2}$ to $-\frac{5}{2}$. The states are $+\frac{5}{2}, +\frac{3}{2}, +\frac{1}{2}, -\frac{1}{2}, -\frac{3}{2}, \frac{-5}{2}$. The different states are due to different angles made by the vectors with each other.

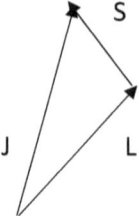

Similarly, there are 4 non-aligned states from $+\frac{3}{2}$ to $-\frac{3}{2}$

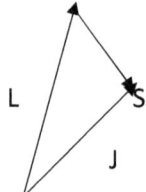

For n=3, l=2, the degeneracy is broken by L-S coupling due to high energy aligned and low energy anti-aligned states.

Let's do the perturbation calculation now.

$V = (-\frac{e}{m}S)(\frac{1}{4\pi\varepsilon_0}\frac{e}{mc^2m^3}L)$

$\approx S \cdot L$

How do we calculate $L \cdot S$?

$J = L + S$

$J^2 = (L+S)(L+S) = L^2 + S^2 - 2L \cdot S$

$L \cdot S = \frac{1}{2}(J^2 - L^2 - S^2)$

$J^2 = j(j+1)$

$L^2 = l(l+1)$

$S^2 = s(s+1)$

The first order energy correction

$E_1 = \langle \Psi_0 | V | \Psi_0 \rangle$

Putting in the value of potential calculated above,

Most of the factors are just numbers, they will come out of integral, the only integral calculation will be $\left\langle \frac{1}{r^3} \right\rangle$

I will skip details on solving it but what we will find is

$E_1 \approx \left(\frac{\alpha}{n}\right)^4 \frac{n}{j+\frac{1}{2}}$

See standard textbooks in the reference section for detailed derivation.

This tells us that j quantum number determines energy splitting.

This means z component of j, determined by number m_j is a good quantum number. The quantum numbers m_l, m_s are of no importance due to L and S coupling as they gave away their relevance to j.

Zeeman Effect

The external magnetic field can affect the spin-orbital coupling. If the field is weak then spin-orbital coupling remains intact and field is a minor perturbation. But if the external field is strong then spin-orbital coupling is broken and the field dominates, with fine structure becoming the perturbation.

Let's first study the weak Zeeman effect. The external magnetic field that is applied to the atom is much weaker than the internal magnetic field generated inside the atom.

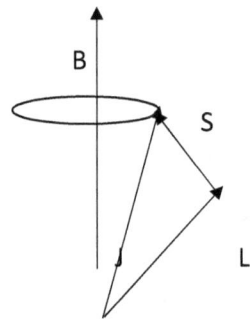

The application of external magnetic field B leads to torque on the vector J which precesses around the magnetic field B. The L and S vectors also precess around J vector.

The magnetic moment of J, $\mu_j = \mu_l + \mu_s = (\frac{-e}{2m}L) + (-\frac{e}{m}s) = \frac{-e}{2m}(L + 2S)$

The perturbation potential $V = -u_j \cdot B = \frac{e}{2m}(L + 2S) \cdot B$

First order energy correction = $\langle \Psi_0 | V | \Psi_0 \rangle$

Or $\frac{e}{2m} B \langle L + 2S \rangle$

The key is to get the average value of L+2S. Note J=L+S. So, J and μ_j(L+2S) are not aligned and form an angle.

$\langle L + 2S \rangle$ can be written as L+S+S=$\langle J + S \rangle$

We want to know average value of $\langle S \rangle$ first

It is projection value of S vector onto J

$$\langle S \rangle = \frac{S \cdot J}{J^2} J$$

$S \cdot J$ as we discussed earlier is nothing but $J^2 + L^2 - S^2$ or $j(j+1) + s(s+1) - l(l+1)$

$s(s+1) = \frac{1}{2}\left(\frac{1}{2}+1\right) = \frac{3}{4}$ as s is fixed at $\frac{1}{2}$.

$$\langle J + S \rangle = \left\langle J + \frac{S \cdot J}{J^2} J \right\rangle = \left\langle \left(1 + \frac{S \cdot J}{J^2}\right) J \right\rangle$$

Putting in value of $S \cdot J$ and J^2

$$\left(1 + \frac{j(j+1) - l(l+1) + \frac{3}{4}}{2j(j+1)}\right) \langle J \rangle$$

The bracket term is the Lande' g factor.

What is $\langle J \rangle$?

If we take magnetic field B to be along z axis, then average value of J will be J_z which is equal to $m_j \hbar$.

$$E_1 = \mu_B g B m_j$$

Here $\mu_B = \frac{e\hbar}{2m}$ is called Bohr magneton.

e.g. for Ground state 1s state, n=1, l= 0, s=$\pm\frac{1}{2}$

j= $|l \pm s| = \left|0 \pm \frac{1}{2}\right| = \frac{1}{2}$ and $m_j = +j$ to $-j = \pm\frac{1}{2}$

$$g = 1 + \frac{\frac{1}{2}\left(\frac{1}{2}+1\right) - 0(1+1) + \frac{3}{4}}{2\frac{1}{2}\left(\frac{1}{2}+1\right)} = 1+1 = 2$$

$$E_1 = \mu_B g B m_j = 2\mu_B B\left(\pm\frac{1}{2}\right)$$

1s orbital is split into two levels +1/2 and -1/2 due to two values of m_j.

Recall that there is no splitting of 1s energy level in Spin-orbital coupling as it does not depend on m_j but only depends on j which is ½.

Strong Zeeman Effect or Paschen-Back Effect

If the external magnetic field is very strong as compared to the internal magnetic field of the atom, L-S coupling is broken. The L and S vectors precess magnetic field B independently.

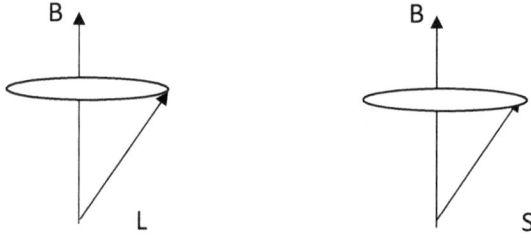

The potential due to orbital angular momentum

$U_L = -\mu_L . B = -(\frac{-e}{2m} L_z). B = \frac{e}{2m} m_l \hbar$ B

The potential due to spin orbital angular momentum

$U_S = -\mu_S . B = -(\frac{-e}{m} S_z). B = \frac{e}{m} m_s \hbar$ B

$U = \frac{e}{2m}(m_l + 2m_s) \hbar$ B

The energy splitting and spectrum will be richer as compared to the weak Zeeman effect as m_l, m_s cause more transitions to happen. m_j is not a good quantum number in the presence of the strong magnetic field unlike the weak effect where m_l, m_s are not good quantum numbers.

Hyperfine Structure

If you are getting tired of fine structure of Hydrogen, there is hyperfine structure to explore as well. We ignored the spin and magnetic moment of proton. If we include the proton, there are new interactions to explore.

The magnetic moment of proton is

$$\mu_p = -\frac{ge}{2m_p} S_p$$

Since the mass of the proton is much heavier than electron, magnetic moment of proton is much smaller than that of electron. That's why we ignored it earlier.

g is called the gyromagnetic ratio.

It is the ratio of magnetic moment to the spin of the proton.

$$g = \frac{\mu}{s}$$

Any moving or spinning charge creates a magnetic field. The gyromagnetic ratio relates the strength of the magnetic field to the spin of the proton that generates the magnetic field. It is determined experimentally. The value of g for proton is 5.56. This value gives a strong evidence that proton is not a point particle, but a composite particle made of quarks.

The value of g for electron is 2.

$$\mu_e = -\frac{ge}{2m_e}S_e = -\frac{e}{m_e}S_e$$

We have two spin vectors and their magnetic moments.

What do you expect?

Obviously, they will couple and combine to form total spin S.

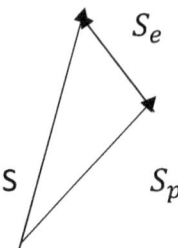

The electron and proton being fermions, their spin is fixed at $\frac{1}{2}$.

If the spins are aligned, then $S = \frac{1}{2} + \frac{1}{2} = 1$

If the spins are opposite, then $S = \frac{1}{2} - \frac{1}{2} = 0$

For spin 1 case, there are three states called triplet states.

Spin of electron = $+\frac{1}{2}$ (↑) or $-\frac{1}{2}$ (↓)

Spin of proton = $+\frac{1}{2}$ (↑) or $-\frac{1}{2}$ (↓)

Triplet states are

↑↑ + ↑↑

↑↓ + ↑↓

↓↓ + ↓↓

They just represent alignment in three different directions

For spin zero, there is one singlet state

↑↓ −↑↓

In the presence of magnetic field, aligned and non-aligned spins will behave differently as the field tries to orient them, leading to split of energy levels.

The perturbation potential

V= −μ · B

The exact expression is a bit tedious as magnetic field generated by proton is not straight forward but for spherically symmetrical case of $l = 0$, the expression simplifies to

$$\langle V \rangle \approx \langle S_p \cdot S_e \rangle$$

This product is easy to calculate as we have seen in LS coupling.

$$S_p \cdot S_e = \frac{1}{2}(S^2 - S_p^2 - S_e^2)$$

$$S_p^2 = S_e^2 = s(s+1) = \frac{1}{2}\left(\frac{1}{2}+1\right) = \frac{3}{4}\hbar^2$$

As you are aware there is $\hbar$ stuck to all spin values and I am not keeping track of all $\hbar$.

What remains to calculate is total spin S^2.

$S = \sqrt{s(s+1)}$, small s quantum number is calculated by adding spins of electron and proton as we did for j quantum number in LS coupling where j= $|l \pm s|$.

Here s= $\left|\frac{1}{2}+\frac{1}{2}\right|=1$ or $\left|\frac{1}{2}-\frac{1}{2}\right|=0$

For triplet states $S^2 = s(s+1) = 1(1+1) = 2\hbar^2$

$S_p \cdot S_e = \frac{1}{2}(2 - \frac{3}{4} - \frac{3}{4}) = +\frac{1}{4}$

For singlet state, $S^2 = 0$

$S_p \cdot S_e = \frac{1}{2}(0 - \frac{3}{4} - \frac{3}{4}) = -\frac{3}{4}$

The energy difference between triplet and singlet states in ground state is $\approx 10^{-6}$ eV.

The higher energy triplet state can flip into the lower energy singlet state, releasing photons with the wavelength of 21cm.

In astronomy, this is called the 21cm line.

It is quite useful in radio astronomy. The radio waves can pass through dust clouds which visible light cannot do. It does take millions of years for the triplet state to flip to singlet state but since Hydrogen is so abundant in galaxies, the flip is observable and is useful to map galaxies.

You guys are expecting intuition behind all these phenomena, right?

Whenever I need to explain obscure and complicated phenomenon, I turn to our venerable politicians.

Each politician carries its own spin. Democrat spin is different than Republican spin. An issue at hand acts like a magnetic field and leads to coupling, call it R-D coupling. If the issue is national security or paying homage to national heroes, Republican and Democrat spins align with magnetic pull of the issue, leading to a lower energy state. If the issue is divisive like immigration or abortion, the spins are anti-aligned, leading to high tension or energy state. The hyperfine structure of congress can lead to unlikely combinations like Red state Democrats couple with moderate Republicans for local issues. We can run out of fine structure coupling in Hydrogen but not the possible couplings in the Congress!

Variational Principle

It is a powerful method when we cannot solve the Schrodinger equation. If we do not solve the Schrodinger equation, we do not have eigenfunctions. If we do not have eigenfunctions then we cannot accurately measure energy levels.

Suppose you have a key but forgot which lock it belongs to.

How will you find it out?

Well, you have to try it on all the locks you have and see which one gets unlocked.

You can do better than that. You start making guesses. You may look at the key and may have clue which kind of lock it belongs to. You check parameters like shape, size, color etc. You make be lucky if you get it right in the first try, otherwise you keep on guessing. This is the essence of variational principle.

The eigenfunctions are like locks. The operator is the key. It operates on the eigenfunctions. Unlocking is the eigenvalue of the eigenfunctions. The key can open up many locks(eigenfunctions). Suppose we are interested in the smallest lock that can be opened. It is the ground state of the eigenfunctions. You need to guess how small the lock can be. You may guess it based on the size of the key. You will get some errors. Keep on trying, going to different stores and see the smallest lock that can be opened with the key. This is the ground state. The obvious problem is that you are never sure how close are you to the ground state.

Operator

Eigenfunction

Let's say the ground state function is something like

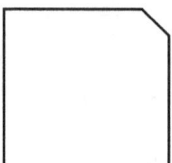

It is the exact eigenfunction of the ground state

$H\Psi_g = E_g\Psi$

You guessed a function like

There is an error here.

If you parametrize the error by α, where $\alpha = \frac{1}{width}$

The width is of the error function.

We want to minimize the error.

We want $\frac{dH}{d\alpha} = 0$

We do not want error function to do anything to the Hamiltonian.

If changing error function is not doing anything to energy values, we are getting close to the true function.

Think of error function as a volume knob

We are fine tuning the knob so that we get the exact volume.

The ground function of a wave usually looks like

Single node

Not like

This is an excited state with many nodes.

We start with the function that looks like the ground state. Mathematically, we will guess a Gaussian function (Bell curve).

How close can we get to the ground state function?

Let's start with any trial function ψ

$$\psi = \Sigma c_n \psi_n$$

It is made of many eigenfunctions with various energies.

What is the average energy?

$$\langle E \rangle = \frac{\langle \psi | H | \psi \rangle}{\langle \psi | \psi \rangle}$$

$\langle \psi | \psi \rangle$ =1 due to normalization condition

$\langle \psi | H | \psi \rangle = c^* \psi^* \lambda_n \, c \, \psi = |c_n|^2 \lambda_n$

$|c_n|^2 \lambda$ means what is the probability ($|c_n|^2$) of each energy level λ_n

e.g. in case of dice, we will have

$\frac{1}{6} \times 1 + \frac{1}{6} \times 2 + \frac{1}{6} \times 3 + \frac{1}{6} \times 4 + \frac{1}{6} \times 5 + \frac{1}{6} \times 6 = 3.5$

$3.5 \geq 1$

So, the average value is always higher or equal to the minimum value possible.

When will average value be equal to the minimum value?

If you are averaging from just one value.

$\langle E \rangle = \langle 1 \times 1 \rangle$

Average of 1 is 1.

Application

The classical application of the variational principle is to find ground state of Helium. The Schrodinger equation for Helium cannot be solved exactly. The hydrogen atom equation is solvable but adding even one electron makes things impossible to solve. The second order differential equations like Schrodinger equation are difficult to solve and adding even one variable can make things very difficult.

Let's write the Hamiltonian for Helium.

H= KE + PE

KE = $\nabla^2 + \nabla^2$

It is made of KE of two electrons. We assume that nucleus is stationary in comparison to electrons.

PE function is more interesting.

PE = electron (1)-proton attraction+ electron (2)-proton attraction+ electron-electron repulsion

Let's write the Hamiltonian in its full glory

$$H = \frac{-\hbar^2}{2m}(\nabla^2 + \nabla^2) - \frac{e^2 2}{4\pi\varepsilon_0 r_1} - \frac{e^2 2}{4\pi\varepsilon_0 r_1} + \frac{e^2}{4\pi\varepsilon_0}\frac{1}{r_1 - r_2}$$

Note here 2 comes from the total charge which is Ze or 2e. Z is atomic number of Helium.

What is the easiest way to calculate ground state of Helium?

We know the ground state of Hydrogen.

$$\psi_H = \frac{4}{\sqrt{\pi a^{\frac{3}{2}}}} e^{-2\frac{r}{a}}$$

Helium is mode of two Hydrogen like atoms, so just combine the two Hydrogen wave functions.

$$\psi_H \psi_H = \frac{8}{\pi a^3} e^{-2\frac{(r_1+r_2)}{a}}$$

Here r_1 and r_2 are the positional coordinates of two electrons and a is the Bohr's radius.

$$a = \frac{4\pi\varepsilon_0}{e^2}\frac{\hbar^2}{m}$$

We know from the study of Hydrogen atom that nth Energy level is

$E_n = \frac{Z^2 E_1}{n^2}$, E_1 is the ground state of Hydrogen with value -13.6 eV

For Helium, Z =2, ground state n=1

So, energy of one hydrogen like atom is $4E_1$

Ground state of Helium= $4E_1 + 4E_1$ =8 × -13.6 = -109 eV

The experimental value is -78.97

This approximation is not close.

What did we miss?

Obviously, the electron-electron repulsion. New interactions happen when two Hydrogen like atoms interact.

We see it all the time in real like. Whether making a cocktail or two people meeting each other, individual behavior is very different than social behavior.

Let's add the contribution of electron-electron repulsion(V_{ee})

This means adding average potential

$$\langle V_{ee} \rangle = \langle \psi | V_{ee} | \psi \rangle = \frac{e^2}{4\pi\varepsilon_0} \int \frac{|\psi|^2}{r_1 - r_2} d^3r_1, d^3r_2$$

Trial function $\psi = \frac{8}{\pi a^3} e^{-2\frac{(r_1+r_2)}{a}}$

Expression becomes

$$\frac{e^2}{4\pi\varepsilon_0} \left(\frac{8}{\pi a^3}\right)^2 \int \frac{e^{-4\frac{(r_1+r_2)}{a}}}{r_1-r_2} d^3r_1, d^3r_2$$

This is a tedious integral. I will spare the details as calculations run a page long.

See the standard textbooks like Griffith or Shankar, as mentioned in the reference section.

Bottom line is

$$\langle V_{ee} \rangle = -\frac{5}{2} E_1 = -\frac{5}{2} \times -13.6 = 34 \text{ eV}$$

Recall that $E_1 = -\frac{m}{2\hbar^2} \left(\frac{e^2}{4\pi\varepsilon_0}\right)^2$

Ground state energy $\langle H \rangle = -109 + 34 = -75$

This is closer to the experimental value.

But it can be refined further. We used nuclear charge Z=2. This is the charge experienced by 2 electrons.

What if effective nuclear charge is different than 2?

The electrons repel each other, and one electron shields the proton from the other electron.

Z can be used as a variable parameter.

We need to replace 2 by Z in our Hamiltonian

The original Hamiltonian, H_0 will become

$$\frac{-\hbar^2}{2m}(\nabla^2 + \nabla^2) - \frac{e^2 Z}{4\pi\varepsilon_0 r_1} - \frac{e^2 Z}{4\pi\varepsilon_0 r_1}$$

$\langle V_{ee} \rangle$ term does not have any Z.

We need to add additional term H_1 to the Hamiltonian to reflect that Z will be different than 2.

$$\frac{e^2}{4\pi\varepsilon_0}\left(\frac{Z-2}{r_1} + \frac{Z-2}{r_1}\right)$$

So total Hamiltonian will be

$$\frac{-\hbar^2}{2m}(\nabla^2 + \nabla^2) - \frac{e^2 Z}{4\pi\varepsilon_0 r_1} - \frac{e^2 Z}{4\pi\varepsilon_0 r_1} + \frac{e^2}{4\pi\varepsilon_0}\left(\frac{Z-2}{r_1} + \frac{Z-2}{r_1}\right) + \frac{e^2}{4\pi\varepsilon_0}\frac{1}{r_1 - r_2}$$

ψ will also need to be changed $\frac{Z^3}{\pi a^3} e^{-Z\frac{(r_1+r_2)}{a}}$

Ground state energy will be

$$\langle H \rangle = \langle \psi|H_0|\psi \rangle + \langle \psi|H_1|\psi \rangle + \langle \psi|V_{ee}|\psi \rangle$$

We need to do more calculations as usual

$\langle \psi|H_0|\psi \rangle = \langle \psi|E_0\psi \rangle = 2E_1 \frac{Z^2}{n^2} = 2E_1 Z^2$ as ($\langle \psi|\psi \rangle = 1$)

$\langle \psi|H_1|\psi \rangle$

Here you will find expression $\langle \psi|\frac{1}{r_1}|\psi \rangle$ and $\langle \psi|\frac{1}{r_2}|\psi \rangle$

ψ is made of two parts with one electron at r_1 and other at r_2.

$\langle \psi\, r_2|\psi\, r_2 \rangle = 1$

$\langle \psi\, r_1|\frac{1}{r_1}|\psi\, r_1 \rangle$ will be left and we call it $\langle \frac{1}{r} \rangle$

This is the average distance of electron from nucleus in one electron state.

Similarly, $\langle \psi\, r_2|\frac{1}{r_1}|\psi\, r_2 \rangle = \langle \frac{1}{r} \rangle$

$$\left\langle \frac{1}{r} \right\rangle = \frac{Z}{a}$$

Lastly $\langle \psi | V_{ee} | \psi \rangle$

Since we skipped integral, I will not go into detail but expression in terms of Z can be calculated.

In terms of Z, after some cleaning up

$$\langle H \rangle = [2Z^2 - 4Z(Z-2) - \frac{5}{4}Z]E_1 = [-2Z^2 + \frac{27}{4}Z]E_1$$

Variational principle says $\frac{dH}{dZ} = 0$

Doing basic differentiation, we get

$$[-4Z + \frac{27}{4}]E_1 = 0$$

Since E_1 is not zero, other term needs to be zero

$$-4Z + \frac{27}{4} = 0$$

$$Z = \frac{27}{16} = 1.69$$

The last step is to put value of Z back into $\langle H \rangle = [-2Z^2 + \frac{27}{4}Z]E_1$

We get $\langle H \rangle = -77.5$ eV

This is even closer to the experimental value of -78.97 eV.

We can only get value greater than or equal to the ground state. If we introduce more parameters, closer match can be achieved.

WKB Approximation

This approximation is useful if potential is changing slowly with distance.

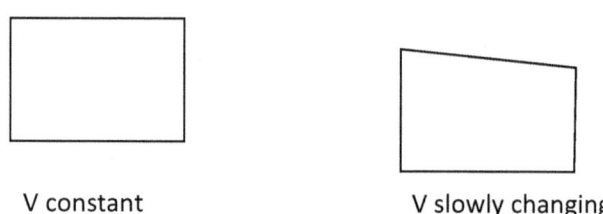

     V constant   V slowly changing

If a barrier is not smooth but varies slowly with distance x, then WKB approximation is useful. Note there is no time dependence of the potential.

We expect wave function of the particle that comes in contact with a barrier to change as well. But if the change in potential is very slow and particle has high frequency then over one wavelength, V is nearly constant.

High frequency state n= high

The high frequency or excited state is called semiclassical state. The ground state would not work with WKB approximation.

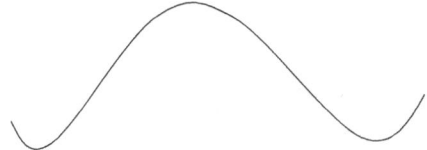

The ground state n=1 is not going to work for WKB approximation.

The wave amplitude A is going to change under the influence of changing potential.

If wave amplitude is constant, then $\frac{\partial A}{\partial x} = 0$ or a straight line on a graph.

But we expect some change in amplitude due to changing potential

$\frac{\partial A}{\partial x} \neq 0$ but changing very slowly

WKB approximation is that second derivative $\frac{\partial^2 A}{\partial x^2} \approx 0$.

This makes sense as first derivative is changing slowly. So, the rate of change of first derivative can be considered zero.

What can we do with $\frac{\partial^2 A}{\partial x^2} \approx 0$.

This assumption will be used when we solve Schrodinger equation.

How can we intuitively think about it?

Let's say you are jogging on a level surface and after a while, the surface becomes uphill.

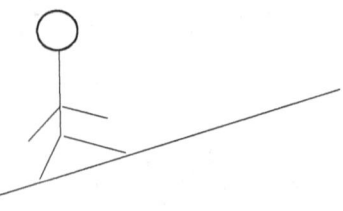

What would happen?

You will slow down; your steps will be less frequent.

This is what happens to a wave.

If the uphill is not too steep, you can still keep jogging but at a slower pace.

This is what WKB approximation predicts.

What if potential is changing fast or as a runner you get a barrier that is steep?

If you come across a wall or a big rock then you will have to jump, right?

This is not jogging. So, your rhythm is going to be distorted quickly. Here WKB approximation will not work.

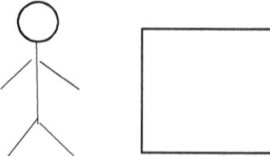

As usual we start with Schrodinger equation

$$-\frac{\hbar^2}{2m}\frac{d^2\psi}{\partial x^2} + V(x)\psi = E\psi$$

$$\frac{d^2\psi}{\partial x^2} = -\frac{p^2}{\hbar^2}\psi$$

Here $p = \sqrt{2m\,E - V}$

P is the momentum here.

The solution to this equation is straight forward and we have seen it many times.

$$\psi = Ae^{i\phi}$$

The phase ϕ is not constant as V is changing with x.

Putting the solution back into the equation, we get

$\frac{d\psi}{dx}$ gives us 2 terms (basic product differentiation)

$\frac{d^2\psi}{\partial x^2}$ gives 4 terms

$$[\partial^2 A + 2i\partial A\partial\phi + iA\partial^2\phi - A(\partial\phi)^2]e^{i\phi} = -\frac{p^2}{\hbar^2}A\,e^{i\phi}$$

To make things easier, separate real and imaginary parts

$2i\partial A\partial\phi + iA\partial^2\phi = 0$ (imaginary)

Dividing by i on both sides and multiplying by A on both sides

$2A\,\partial A\partial\phi + A^2\partial^2\phi = 0$

It is equivalent to $\partial(A^2\partial\phi) = 0$

To see this, differentiate $A^2\partial\phi$

This is basic product differentiation of 3 terms A A $\partial\phi$

$AA\partial^2\phi + A\,\partial A\partial\phi + A\,\partial A\partial\phi = 0$

Or $2A\,\partial A\partial\phi + A^2\partial^2\phi = 0$

Now ∂ constant $=0$

This means

$A^2\partial\phi$ = constant

$A = \dfrac{C}{\sqrt{\partial\phi}}$

Going back to the real part

$\partial^2 A - A(\partial\phi)^2 = -\dfrac{p^2}{\hbar^2}A$

$\partial^2 A$ term can be ignored as this is the premise of WKB approximation.

$(\partial\phi)^2 = \dfrac{p^2}{\hbar^2}$

$\dfrac{\partial\phi}{\partial x} = \pm\dfrac{p}{\hbar}$

$\phi = \pm\dfrac{1}{\hbar}\int P dx + K$ (integration constant)

Putting value of $\partial\phi$ in the equation for amplitude A

$A = \dfrac{C}{\sqrt{\partial\phi}} = \dfrac{C}{\sqrt{\dfrac{p}{\hbar}}} = \dfrac{C\sqrt{\hbar}}{\sqrt{p}}$

$\psi = Ae^{i\phi} = \dfrac{C\sqrt{\hbar}}{\sqrt{p}}\,e^{\frac{i}{\hbar}\int Pdx+K}$

$e^{\frac{i}{\hbar}\int Pdx+K}$ can be written as $e^{\frac{i}{\hbar}\int Pdx}\,e^{iK}$

$\psi = Ae^{i\phi} = \dfrac{C\sqrt{\hbar}}{\sqrt{p}}\,e^{\frac{i}{\hbar}\int Pdx}\,e^{iK}$

$C\sqrt{\hbar}\, e^{iK}$ can be absorbed into another constant, we will again call it C.

$$\psi = \frac{C}{\sqrt{p}}\, e^{\pm\frac{i}{\hbar}\int P\, dx}$$

This equation makes sense as probability of finding a particle $|\psi|^2 \approx \frac{1}{p}$

So, the particle is least likely to be found in areas where it is moving fast.

Also note $p = \sqrt{2m\, E - V}$

If V is increasing, p will decrease. It is like saying that moving uphill slows you down.

Case of V > E

This is like coming across a mountain. It's height or potential is too high for you to throw a rock across it.

P is negative in this case. So, we take its magnitude $|p|$

The magnitude will replace p in the formula for ψ.

$$\psi = \frac{C}{\sqrt{|p|}}\, e^{\pm\frac{1}{\hbar}\int |p|\, dx}$$

+ and − sign in front of potential represent left and right moving waves. The function is real as there is no i in it. The reason is that it is a real decaying exponential as it should be if potential is high. The oscillating function can also be written in exponential form and it will have the form e^{-ikx} as it is an imaginary exponential as in reality it is oscillating not decaying.

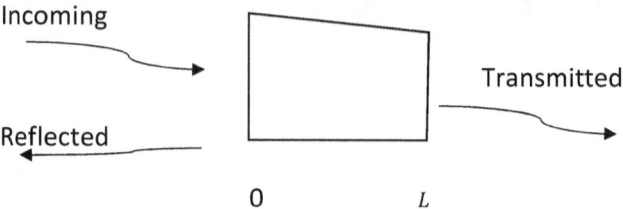

Transmission depends on $\frac{\psi_T}{\psi_I} \approx e^{-\frac{1}{\hbar}\int |p|dx}$

Probability of transmission $|\frac{\psi_T}{\psi_I}|^2 \approx e^{-2\gamma}$ where $\gamma = e^{\frac{1}{\hbar}\int_0^L |p|dx}$

Classic case is Alpha particle

Alpha particle is Helium nucleus with 2 protons and neutrons. It is stuck inside Uranium -235 nucleus. It is being repelled by positive charged nucleus, but it is also held back by a powerful barrier of nuclear force which is attractive in nature.

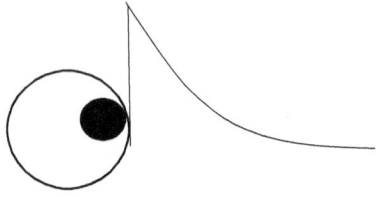

Probability of emission of alpha particle is $e^{-2\gamma}$.

To increase probability, γ has to be small. This means $\int_0^L |p|dx$ should be small.

$p \approx \sqrt{V - E}$

So alpha particle needs have high energy, potential barrier has to be small and tunnel length (0 to L) should also be small. This makes sense.

The probability is small but alpha particle gets emitted as it keeps on bombarding the barrier and once in a while it gets through.

You may be wondering that anything is possible in quantum mechanics. Even if barrier is very high and energy low, things still happen.

What's the probability of throwing a ball into space?

According to classical mechanics, it is impossible. Escape velocity is too high, and your arm does not have the strength to give that velocity to the ball. But quantum mechanics says, it can still tunnel through.

Which is correct?

Let's do a rough calculation and see what we get.

If we take 1 kg ball, it will face the gravitational potential energy barrier

$V = mgh = 1 \times 10 \times 12000 = 12 \times 10^4$

Here we took height of atmosphere at 12 km. We took g to be constant for simplicity even though it decreases with height.

KE of the ball if we throw at 100 km/hr $\approx 30 m/s$

KE = $\frac{1}{2} mv^2 = \frac{1}{2} \times 1 \times (30)^2 = 450$

$p \approx \sqrt{V - E} \approx 10^2$, $\hbar \approx 10^{-34}$

$\gamma = e^{\frac{1}{\hbar} \int_0^L |p| dx} \approx \frac{10^2}{10^{-34}} = 10^{36}$

Probability of transmission $\approx e^{-2\gamma} = e^{-2 \times 10^{36}}$

Number of attempts = time taken for each attempt × frequency of event= τf

If we throw 1 ball each second, then $f = 1$

Number of attempts × probability of each attempt = 1 (at least 1 ball has to reach moon)

$1 \times \tau \times e^{-2\times 10^{36}} = 1$

$\tau = e^{2\times 10^{36}}$ seconds

This is more than the age of the universe.

You just need to be patient!

Case of V= E

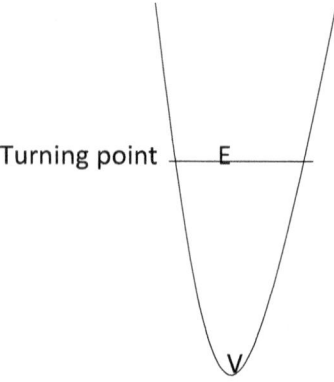

Turning point — E

V

When V= E, it is called a turning point. Think of a pendulum, KE= PE at turning point where pendulum turns back.

It is a tricky point mathematically.

$P = \sqrt{V - E} = 0$

$\psi = \frac{1}{\sqrt{p}} = \infty$

$\lambda = \frac{2\pi}{p} = \infty$

WKB approximation fails here.

What should we do?

There is a gap between V > E and V < E. V=0 is like a cliff in between these two regions.

We need to build a bridge that connects two regions.

It is called patching function. It will connect two well defined regions. Once things are connected, we can move across bridge and patching function will no longer be needed.

Let's look at right turning point.

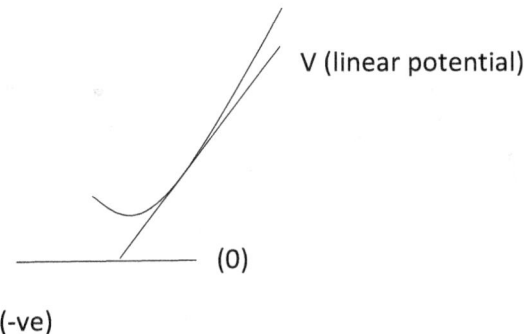

At the turning point, we make the approximation that V is a straight line and then add small change to it.

$$V(x) = V(0) + \frac{dV}{dx}x$$

$\frac{dV}{dx}x$ = how it changes with x.

Taking case of $\frac{dV}{dx} > 0$

Schrodinger equation becomes

$$-\frac{\hbar^2}{2m}\frac{d^2\psi}{\partial x^2} + (V(0) + \frac{dV}{dx}x)\psi = E\psi$$

$$\frac{d^2\psi}{\partial x^2} = -\frac{2m}{\hbar^2}(E - V(0) - \frac{dV}{dx}x)\psi$$

At turning point V=E

$$E - E - \frac{dV}{dx}x$$

$$\frac{d^2\psi}{\partial x^2} = -\frac{2m}{\hbar^2}\left(-\frac{dV}{dx}x\right)\psi$$

Defining $\alpha = \frac{2m}{\hbar^2}\left(\frac{dV}{dx}\right)$

$$\frac{d^2\psi}{\partial x^2} = \alpha x \psi$$

Since α has dimensions of inverse length, we define a new variable which is dimensionless.

$$z = \alpha x$$

This is a common trick.

Replace the x variable with z.

$$\frac{d^2\psi}{\partial z^2} = z\psi$$

$$\frac{d^2\psi}{dz^2} - z\psi = 0$$

This is like $\frac{d^2y}{dx^2} - xy = 0$

The reason we did these tricks is to get the equation in this form as solution is well known in mathematics.

The solution is power series.

$$y = \Sigma a_n x^n$$

It says our house(y) is made of many bricks.

xy term is $\sum_{n=0}^{\infty} a_n x^{n+1}$

Doing differentiation on y twice, we get

$$\frac{d^2y}{dx^2} = \sum_{n=2}^{\infty} a_n n(n-1) x^{n-2}$$

$$\sum_{n=2}^{\infty} a_n n(n-1) x^{n-2} - \sum_{n=0}^{\infty} a_n x^{n+1} = 0$$

This is an algebraic equation and can be solved easily.

We get recurrence relation.

There are two series

$$a_0 = 1 + \frac{x^3}{()} + \frac{x^6}{()} + \frac{x^9}{()}$$

$$a_1 = x + \frac{x^4}{()} + \frac{x^7}{()} + \frac{x^{10}}{()}$$

There are some other factors in the equations which for simplicity I won't write.

This is like saying you have 2 kind of bricks and each brick has a formula (recurrence relation). Each brick determines the next one and so on so that they fit exactly to form the house.

If you plot these two solutions, you will get

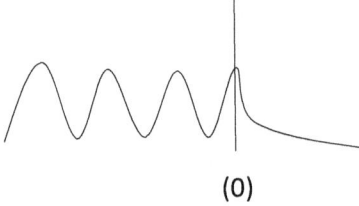

(0)

Oscillating   Decaying exponential

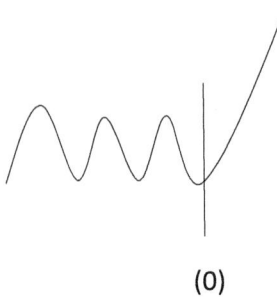

(0)

Oscillating   Increasing exponential

These are called Airy Solutions

$$\psi_z = a A_{iry} + b B_{iry}$$

Or

$$\psi_p \approx a e^{\frac{-2}{3}(\alpha x)^{\frac{3}{2}}} + b e^{\frac{2}{3}(\alpha x)^{\frac{3}{2}}}$$

We need this to match left and right-side solutions using our bridge function.

$$\psi_r = \frac{1}{\sqrt{|p|}} \left(C e^{+\frac{1}{\hbar}\int |p| dx} + D e^{-\frac{1}{\hbar}\int |p| dx} \right)$$

Right side function is like facing a barrier with V> E. We expect a decaying function. This is what we have. We will ignore C part later as it is not a decaying function.

At V= E

$$p \approx \sqrt{2m(E - E - \frac{dV}{dx}x)} = \hbar\alpha^{\frac{3}{2}}\sqrt{-x}$$

$$\psi_r = C e^{\frac{-2}{3}(\alpha x)^{\frac{3}{2}}} + D e^{\frac{2}{3}(\alpha x)^{\frac{3}{2}}}$$

Match the functions

$$\begin{matrix} a \\ b \end{matrix} () = () \begin{matrix} C \\ D \end{matrix}$$

( ) some factors

Similarly, left side function is an oscillating function as it is inside the well.

$$\psi_l = \frac{1}{\sqrt{p}} \left(A e^{+\frac{i}{\hbar}\int p dx} + B e^{-\frac{i}{\hbar}\int p dx} \right)$$

Doing same tricks, we match it with patching function

$$\begin{matrix} a \\ b \end{matrix} () = () \begin{matrix} A \\ B \end{matrix}$$

Doing more algebraic calculations, we get rid of the middleman patching function and match, see standards texts in the reference section if you want the whole deal.

$$\begin{smallmatrix}C\\D\end{smallmatrix}() = ()\begin{smallmatrix}A\\B\end{smallmatrix}$$

C function is zero as it blows up

$A = -iDe^{i\frac{\pi}{4}}$ $B = iDe^{-i\frac{\pi}{4}}$

Putting these values in the oscillating left sided function we get

$\psi_l = A\, e^{+\frac{i}{\hbar}\int p dx} + B e^{-\frac{i}{\hbar}\int p dx}$

$\psi_l = -i\, De^{i\frac{\pi}{4}}\, e^{+\frac{i}{\hbar}\int p dx} + i\, De^{-i\frac{\pi}{4}}\, e^{-\frac{i}{\hbar}\int p dx}$

$\psi_l = -i\, D(e^{i\frac{\pi}{4}}\, e^{+\frac{i}{\hbar}\int p dx} - e^{-i\frac{\pi}{4}}\, e^{-\frac{i}{\hbar}\int p dx})$

Here you will find $-iD(e^{i\theta} - e^{-i\theta})$

Using identity $\sin\theta = \dfrac{e^{i\theta} - e^{-i\theta}}{2i}$

We get $2D \sin\theta$

OR $\psi_l = \dfrac{2D}{\sqrt{p}} \sin\left(\dfrac{1}{\hbar}\int P dx + \dfrac{\pi}{4}\right)$

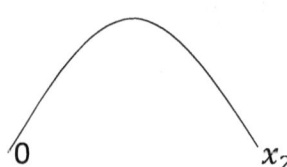

The function ψ_l should end at turning point which we will now call at x_2

The only way for it to send is if $\sin 0$ or $n\pi$

$\dfrac{1}{\hbar}\int P dx + \dfrac{\pi}{4} = n\pi$

$\int P dx = \left(n - \frac{1}{4}\right)\pi\hbar$

Solving equations is fine but what is going on here?

What does adding $\frac{\pi}{4}$ means?

Ok, we have an oscillating function inside the well and a decaying function outside. They don't match at the turning point. They are short by $\frac{\pi}{4}$. We sought the help of middleman.

The middleman said if you want to meet at the turning point, you need to extend the hand of friendship!

$\frac{\pi}{4}$ is like this extension of hand so that oscillating and decaying function can meet at the turning point.

Half harmonic oscillator

Let's solve the case of half the harmonic oscillator or pendulum, from 0 to x_2

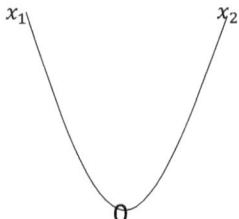

The potential of the oscillator is $\frac{1}{2}m\omega^2 x_2^2$

Putting value of potential to find p

$p = \sqrt{2m(E - V)}$

And then doing integration

$\int_0^{x_2} P dx$

We will find this integral is $\frac{\pi E}{2w}$

$\frac{\pi E}{2w} = \left(n - \frac{1}{4}\right) \pi \hbar$

$E = \left(2n - \frac{1}{2}\right) \hbar w$

This is exact solution for energy quantization.

Similar thing can be done to the left side harmonic oscillator, from 0 to x_1.

Solution is similar as well

$\sin \left(\frac{1}{\hbar} \int P dx + \frac{\pi}{4}\right)$

We will use identity $-\sin(-\theta) = \sin\theta$

So, we can write $-\frac{1}{\hbar} \int P dx - \frac{\pi}{4}$

Let's combine two half harmonic oscillators into a single full oscillator.

The region of turning point on either side is equivalent and is represented by two solutions.

So, $\sin(x_2) = \sin(x_1) + n\pi$

$\int_0^{x_2} P dx + \frac{\pi}{4} = -\int_{x_1}^{0} P dx - \frac{\pi}{4} + n\pi$

$\int_0^{x_2} P dx + \int_{x_1}^{0} P dx = \left(n - \frac{1}{2}\right) \pi \hbar$

$\int_{x_1}^{x_2} P dx = \left(n - \frac{1}{2}\right) \pi \hbar$

Using the process of putting the formula of harmonic oscillator $\frac{1}{2} m\omega^2 x^2$

into $p = \sqrt{2m(E - V)}$ and carrying put the integration of $\int_{x_1}^{x_2} P dx$, we get

$$E = \left(n - \frac{1}{2}\right)\hbar w$$

This is the exact result for the full harmonic oscillator.

Note here that simply putting value of p and doing $\int P dx$ integral, we solved the energy levels without need of solving the Schrodinger equation.

This is part of old quantum theory and is sometimes referred as Bohr -Sommerfeld quantum rule.

Time Dependent Perturbation Theory

The perturbation which is time dependent, is obviously going to complicate things.

$$\Psi_t = \Sigma c_n \Psi_n e^{\theta}$$

e^{θ} does not contribute to probability as it cancels out with the complex conjugate part $e^{-\theta}$. If someone jumps around, it does not change the probability of where that person is.

Here c_n is not time dependent. But if V is time dependent then c_n will also become time dependent.

Let's calculate time dependence of probability coefficients, c_n

$$i\hbar \frac{\partial \psi}{\partial t} = H\psi$$

$$i\hbar \frac{\partial (\Sigma c_n \psi_n e^{\theta})}{\partial t} = (H_0 + V)\Sigma c_n \psi_n e^{\theta}$$

Using product differentiation rules

$$i\hbar \frac{\partial (\Sigma c_n)}{\partial t} \psi_n e^\theta + i\hbar \frac{\partial (e^\theta)}{\partial t} \Sigma c_n \psi_n = (E_0 + V) \Sigma c_n \psi_n e^\theta$$

Now $\theta = -\frac{i}{\hbar} \int E \, dt$

Differentiating it yields $E_0 \psi$ which gets cancelled with similar term on the right side, so remaining terms are

$$i\hbar \frac{\partial (\Sigma c_n)}{\partial t} |\psi_n\rangle e^\theta = V \Sigma c_n |\psi_n\rangle e^\theta$$

Taking inner product with $\langle \psi_m | e^{\theta^*}$

Where $\theta^* = \frac{i}{\hbar} \int E \, dt$

It will simplify things on the left side as inner product is 1 if m=n and zero if m≠n.

The right-hand side cannot use this trick as there is V stuck in between.

Eliminating terms after inner product we get

$$i\hbar \frac{\partial (\Sigma c_n)}{\partial t} = \Sigma c_n \langle \psi_m | V | \psi_n \rangle e^\theta e^{\theta^*}$$

Combining θ terms into omega (w) term $w = \frac{E_m - E_n}{\hbar}$

$$i\hbar \frac{\partial c_m}{\partial t} = \Sigma V_{mn} e^{i w_{mn} t} c_n$$

If a state goes from initial to final configuration after time t, we have

$$C_f = \frac{1}{i\hbar} \int_0^t V_{mn} e^{i w_{mn} t'} c_n \, dt'$$

And probability is $|c_f|^2$

Note here t is time taken from initial to final state and t' is the time dependence of the potential V. I will not always keep this distinction moving forward for the sake of simplicity.

c_n can be eliminated in the first order approximation.

If initial state is n and if it remains there forever, then

Zeroth approximation will be $c_n = 1$ and $c_m = 0$

In the first order approximation, if we calculate $\frac{\partial c_n}{\partial t}$ and put zeroth order values

$\frac{\partial c_n}{\partial t} = 0$ as $c_m = 0$

This means c_n is a constant term, with value one and it will simplify the first order equation to

$C_f = \frac{1}{i\hbar} \int_0^t V_{mn} e^{iw_{mn}t'} \, dt'$ where ($c_n = 1$)

Things are never straight forward. There may be intermediate states between initial and final states. It is like saying you can jump from the first floor to the ground floor or take stairs where several intermediate steps are involved. To account for intermediate steps, we need higher order term to describe the process fully.

i→n→m

$i\hbar \frac{\partial c_m}{\partial t} = \sum V_{mi} e^{iw_{mi}t} c_i + \sum V_{mn} e^{iw_{mn}t} c_n$ (second order)

The value of C_n can be substituted in the second order term

$C_n = \sum V_{ni} e^{iw_{ni}t}$

$C_m(second\ order) = -\frac{1}{\hbar^2} \int V_{mn} e^{iw_{mn}t''} \, dt'' \int V_{ni} e^{iw_{ni}t'} \, dt'$

We could keep adding more higher order terms but for us second order term is good enough.

This is too much math, let's go to the world of analogy. When we throw a die, probability of getting 1,2,3,4,5 and 6 numbers is $\frac{1}{6}$.

Here $|C|^2 = \frac{1}{6}$ and $C_n = \frac{1}{\sqrt{6}}$

$\psi = C_n e^\theta$

The probability of each number does not change with time as the shape of die is not changing. You can think of e^θ as how high or low you throw the die. It does not affect the probability.

What if the shape of die is changing. May be the thrower is physically very strong and is crushing and changing the shape of die each time he throws it. Here shape will not be constant, V is time dependent and $C_n \neq \frac{1}{\sqrt{6}}$

$C_n = \Sigma V_{nm} e^\theta C_m$

C_n is the probability of getting, say number 2.

C_m is the probability you got by throwing the die earlier, say number 1.

V_{nm} is how strongly you bend the die and change its shape.

There you have it, the venerable Time independent perturbation theory.

It wasn't too bad!

Oscillating Potential

This special case is important as electro-magnetic radiation is made of fluctuating electric and magnetic fields and it is useful to see how an oscillating perturbation affects the probability of the wave function.

Oscillating potential = $V \cos wt$

Let's put this in our formula of probability coefficient

$C_f = \frac{1}{i\hbar} V_{fi} \int_0^t \cos wt' \, e^{iw_{fi}t'} \, dt'$

We will focus on frequency omega, w

Using the identity

$$\cos\theta = \frac{e^{i\theta}+e^{-i\theta}}{2}$$

$$\cos wt = \frac{e^{iwt'}+e^{-iwt'}}{2}$$

Relabeling $e^{iw_f it'} = e^{iw_0 t'}$

And adding this to the potential we get

$$C_f = \frac{1}{2i\hbar} V_{fi} \int_0^t [e^{i(w_0+w)t'} + e^{i(w_0-w)t'}]dt'$$

Doing integration by putting limits (t and 0) into the equation and remember $e^0 = 1$, we have

$$\frac{e^{i(w_0+w)t}-1}{w_0+w} + \frac{e^{i(w_0-w)t}-1}{w_0-w}$$

If we chose a special case where oscillating potential w has very similar energy to the energy difference between initial and final states $w_0(E_f - E_i)$, the equation can be further simplified

If w_0 is very close to w, then $w_0 + w \ggg w_0 - w$

e.g. if $w_0 = 10^5$ and $w = 10^4$

$w_0 + w = 10^9$ and $w_0 - w = 10^1$

$10^9 \ggg 10^1$

$w_0 + w$ is a very big number and it being in the denominator

$\frac{e^{i(w_0+w)t}-1}{w_0+w}$ term is very small, and we can safely ignore it.

Only $\frac{e^{i(w_0-w)t}-1}{w_0-w}$ term is left.

This term can be written as

$$\frac{e^{i(w_0-w)\frac{t}{2}}}{w_0-w} [e^{i(w_0-w)\frac{t}{2}} - e^{-i(w_0-w)\frac{t}{2}}]$$

The reason we wrote this expression is to use the formula

$$\sin\theta = \frac{e^{i\theta} - e^{-i\theta}}{2i}$$

Thus $\frac{e^{i(w_0-w)\frac{t}{2}}}{w_0-w} \sin[(w_0-w)\frac{t}{2}]$

$$C_f = \frac{1}{i\hbar} V_{fi} \frac{e^{i(w_0-w)\frac{t}{2}}}{w_0-w} \sin\left[(w_0-w)\frac{t}{2}\right]$$

The probability $|C_f|^2 \approx \sin^2$

The upshot is that oscillatory potential leads to transition probability which is also oscillatory.

Have you played with yo-yo ball?

You give oscillatory energy to the ball with your hand and the ball moves up and down, also getting oscillatory movements.

Note $\int_0^t [e^{i(w_0+w)t'} + e^{i(w_0-w)t'}] dt'$ has average value of zero.

e.g. if $w_0 = 1, w = 2$

then exponents will be $e^{3t} + e^{-1t} = e^{2t}$

But if we had chosen negative values $w_0 = -1, w = -2$

Then exponent value would be e^{-2t}.

So positive and negative values will cause the average to be zero.

This is not what we want as we want transition probability to be maximum.

To achieve it, one of exponents has to be zero.

$w_0 + w = 0$ or $w = -w_0$

Since $w_0 = \frac{E_f - E_i}{\hbar}$

$E_i = E_f + \hbar\omega$

This is the energy conservation for emission where $\hbar\omega$ is emitted as radiation.

Similarly, for $w_0 - w = 0$

$E_i = E_f - \hbar\omega$

This is the energy conservation for absorption.

The above expressions can be used in transition probability formula and replace sin^2

Transition probability P = $|c_f|^2 = \frac{2\pi V_{fi}^2}{\hbar} \delta(E_f - E_i + \hbar\omega)t$

δ function here means the condition that if you supply or take out exactly $\hbar\omega$ then transition probability is maximized.

Golfers know this, a precise shot or energy is needed to put the golf ball in the hole. Any more or less energy is not going to cut it.

Transition rate $\Gamma = \frac{dP}{dt} = \frac{2\pi V_{fi}^2}{\hbar} \delta(E_f - E_i + \hbar\omega)$ Or $\delta(E_f - E_i - \hbar\omega)$

$\Gamma = \frac{2\pi V_{fi}^2}{\hbar} \rho_f$

ρ_f is density of final states and is useful to use if there are many states that a particle can go into.

This expression is called Fermi's Golden Rule.

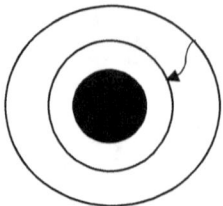

The excited state can go to the ground state by emitting photon with energy $\hbar\omega$.

The ground state can go to the excided state by absorbing photon with energy $\hbar\omega$.

The golden fermi rule is useful when there are lot of final states that are available.

It is like giving energy to rocket to get into space. The initial state of earth can go into endless states in space by giving precise energy so that escape velocity can be achieved.

Selection Rules

The transitions between different states or orbits obey certain rules to make our life easier.

If an electron makes a transition between one orbit to another, it can emit a photon which carries away 1 unit of angular momentum.

So, in order to preserve angular momentum, L changes by 1 unit.

Change in L is ± 1 after transition.

Parity is also conserved in a process. The parity can be either +1 or -1.

It is a multiplicative number. The photon has parity of -1.

Parity $\pi_i = \pi_f (-1)$

To conserve parity the final and initial states need to have opposite parity.

What about z component of angular momentum.

The emitting photon may not have any angular momentum along the z direction or could have 1 unit in the z direction.

So, quantum number $m = 0, \pm 1$

e.g. Can Hydrogen H_{210} (n=2, l=0, m=0) transition into H_{211} ?

Answer is No. The l did not change in this transition. The l has to be ± 1. The transition is forbidden.

What about transition to H_{200}?

This is allowed as change in l is -1.

Parity

Let's do a short review of parity.

Parity is like mirror symmetry. Your mirror image has special components in the opposite direction.

Parity changes x, y and z into -x, -y and -z.

Physics should not change by change of coordinates.

A glass of water is glass of water in the mirror as well and if released, both will fall with same speed as they hit the ground.

Some things change, some do not in the mirror.

The parity can , ± 1.

The color of your shirt will not change, parity is +1.

The left hand will become right hand in the mirror, parity is -1.

The parity operator is π.

In spherical coordinates, parity will cause

$r \to r$ (distance from origin is unchanged)

$\theta \to \pi - \theta$

$\phi \to \pi + \phi$

To calculate parity, we have to use change of coordinates in the solution of angular component of Hydrogen atom or spherical harmonics Y (θ, ϕ).

Solving the full spherical harmonics equation is tedious but

Choosing a special case where l = m, we have

Y (θ, ϕ) $\sim e^{il\phi} \sin\theta$

Note here $e^{im\phi} = e^{il\phi}$ in our special case.

The $\sin\theta$ comes from table of spherical harmonics. If you look at the table where l = m, you will find $\sin\theta$.

$e^{il\phi} \sin\theta$ after change of $\theta \to \pi - \theta$ and $\phi \to \pi + \phi$

$e^{il\phi} e^{i\pi\phi} \sin(\pi - \theta)$

Using identities $\sin(\pi - \theta) = \sin\theta$ and $e^{i\pi} = -1$

We have result $(-1)^l$ Y (θ, ϕ)

So, Parity formula is $\pi = (-1)^l$

Thus, parity only depends on quantum number l .

The elementary particles have intrinsic parity. This comes from Quantum Field Theory which we will learn later.

You can think of particles as people, we have parity that's why we have a mirror image.

By convention, particles like electrons, protons and neutrons have parity of +1.

The anti-particles have parity of -1.

Electric Dipole Approximation

The selection rules of l and m we have studied thus far apply to a certain approximation. It is applicable to one electron jump and interaction with classic electric field only. The effect of magnetic field and multiple electron jumps will complicate the picture and selection rules need to be modified. The electric dipole transitions are far more common than the magnetic poles. If we do calculation with magnetic fields, the transition probability will be small. I won't bother with less common magnetic and multiple transitions. Let's concentrate on the dominant electric dipole transitions.

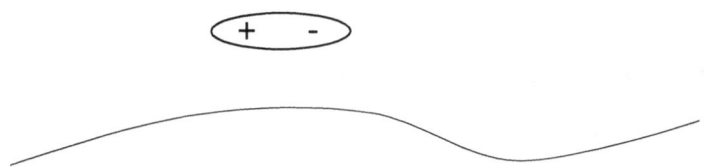

The proton and electron form an electric dipole. The electric field is considered to vary slowly in vicinity of dipole and can be considered constant.

$$\vec{E} = E\cos(kr - wt)$$

For E to be constant, we want $e^{ikr} \approx 1$

Or $kr \approx 0$

$\frac{2\pi}{\lambda} r \approx 0$

$\lambda \gg r$

Now, $V_{fi} \approx \langle f | \varepsilon . r | i \rangle$

ε is unit of polarization and represents direction of E. There are other factors in the derivation which are not important for our purpose.

This is the electric dipole approximation. We are interested in transitions in this approximation.

What do we do with $\langle f | \varepsilon . r | i \rangle$?

The expression can be used to derive selection rules for l and m quantum numbers.

We start with commutation relation

$[L_z, z] = 0$

If we take its expectation value, it should also be zero. So, let's take it

$\langle n'l'm' | [L_z, z] | nlm \rangle = 0$

$[L_z, z] = L_z z - z L_z = (m'\hbar)z - z(m\hbar)$, putting it back into equation

$\langle n'l'm' | [(m'\hbar)z - z(m\hbar)] | nlm \rangle$

Or $(m' - m)\hbar \langle n'l'm' | z | nlm \rangle = 0$

This is the expression we were looking for. We have the electric dipole approximation and m value in the expression.

One of the terms has to be zero. Obviously, dipole approximation is not zero.

So $m' - m = 0$ or $m' = m$.

Similar trick can be used for other commutation relation $[L_z, x] = i\hbar y$ and $[L_z, y] = -i\hbar x$.

After some algebraic manipulation, you will find the condition $(m' - m)^2 = 1$

Or $\Delta m = 1$

The selection rule for l involves using commutation expression $[L^2, [L^2, r]]$

The derivation is rather tedious, and I will not go into detail as there is no new insight gained. The selection rule is again $\Delta l = 1$

Adiabatic approximation

Wave functions depend on the potential of the problem.

What if potential is changing with time.

What will happen to a wave function e.g. will a ground state function remain in ground state or becoming something else.

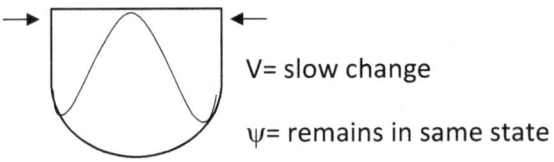
V= slow change
ψ= remains in same state

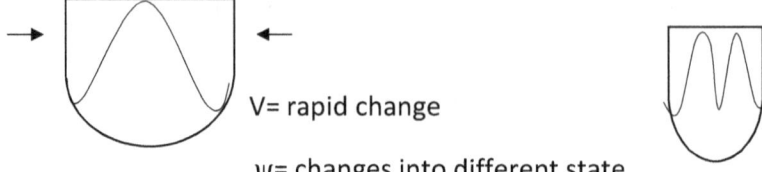
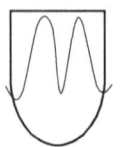

V= rapid change

ψ= changes into different state

It intuitively makes sense. You can think of a person juggler playing with balls in a rhythm. If he moves very slowly, he will maintain his rhythm, but if he starts running, he will lose control of the rhythm.

Time (moving of juggler) >>> Time (juggling of balls)

Mathematically we have to prove

$|\langle \psi_{v,slowchange} | \psi_{final} \rangle|^2 = 0$, if both states are different and 1 if both states are same.

There is lot of confusing terminology in perturbation theory. I am going to give an example of golf and then go through the tedious process of proving the above assertion.

You are a golfer. You know how to adapt to conditions.

ψ_g is wave function. It represents a shot that goes into the hole from a given distance.

What if there is a wind blowing?

Obviously, you have to adapt your shot to the conditions.

$$\psi_w = \Psi_g + \sum \frac{V_{kg}}{E_g - E_k} \Psi_k$$

The above equation tells us that new wave function with wind present, will be the original function + correction made due to wind.

The correction will depend on the perturbation potential V as it modifies how much energy you have to put into the shot and any change in your position.

This is basic time independent perturbation theory.

What if wind changes with time?

In this case, we have to add a time dependent function to the potential.

$Vf(t)$

$\Psi_t = \sum c_n \Psi_n e^{\theta}$

If V is independent of time, then e^{θ} cancels out with complex conjugate part and does not contribute to the probability. It is like saying that blinking of eyes with time, does not contribute to your shot as a golfer.

But if V changes with time like wind is getting stronger then c_n will be time dependent.

$C_n(t) = 1 - \frac{i}{h} V_{nn} \int f \, dt$

c_n is like probability of getting the shot directly into the hole. Let's say probability of getting shot into the golf hole is 70 %. But this probability has to be modified if wind is becoming stronger. This is what this equation does. Here 1 represents 70% probability and you add correction to it.

c_n is just one type of shot. There are modifications needed for other type of shots too.

Second order $C_m = -\frac{i}{h} V_{mn} \int f e^{\theta_1}$

The adiabatic expansion is used when you try to solve the above integrals, the term with $\frac{df}{dt}$ is ignored as it is changing very slowly.

Total $\Psi(t) = 1 - \frac{i}{\hbar} V_{nn} \int f dt$ (first order) $+ \left(-\frac{i}{\hbar} V_{mn} \int f e^{\theta_1}\right)$ second order $\times \psi e^{\theta}$

$\psi_w = \Psi_g + \sum \frac{V_{kg}}{E_g - E_k} \Psi_k$

So, calculating $|\langle \psi_t | \psi_w \rangle|^2$ by putting the above equations, we find it will be zero if both states are different and one if both states are same.

This is the derivation. I tried to simplify it, but it is still tedious and messy, but I hope you got the zest if it.

Moreover, I don't have to tell you that golf is being used as an analogy only. Do not take above examples literally. No one has ever played golf using adiabatic approximation!

Geometric Phase

We have seen that if potential is changing slowly (adiabatic approximation) then eigenfunctions pick up a time dependent phase

$C_n \sim e^{-\frac{i}{\hbar} \int E dt}$ or e^{θ}

This is called the dynamic phase

There is another phase that is picked up, called the geometric phase.

ψ_0 after time t (in a slowly changing V) $\to \psi_0$, remains unchanged

$$\psi_0(t) = \Psi_0 e^{\theta} e^{i\gamma n}$$

$e^{i\gamma n}$ is called the Geometric phase.

What does it mean in simple English?

Let' say we have a performer juggling balls. He has a style or rhythm to the juggling.

He wants to impress the audience. So, he starts moving in a loop trying to maintain his rhythm. He knows he has to move very slowly (adiabatic approximation) to avoid his rhythm getting out of control.

He moves in a special way called parallel transport. The vector length and direction are kept the same as he moves about the loop. In General Relativity parallel transport shows curvature of space.

Change of direction is not allowed.

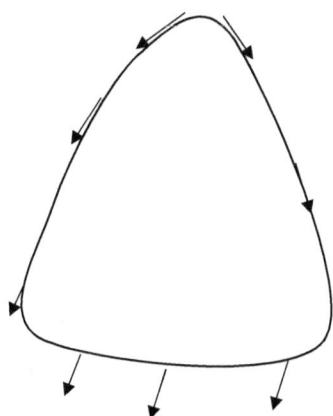

As you finish going around the loop, you will find that the last vector will be at an angle with respect to the starting vector.

θ is the geometric phase angle.

So, our performer will be facing in a different direction as he finishes the loop.

Does it matter?

Of course, if you are the camera person and the performer is facing in a different direction on finishing the loop, you have to adjust the camera to take the picture.

Dynamic phase e^θ represents how long was your trip.

Geometric phase $e^{i\gamma}$ represents where did you go for the trip.

$$\psi_0(t) = \Psi_0 e^\theta e^{i\gamma n}$$

Let's put the value of wave function in the Schrodinger equation.

$$i\hbar \frac{\partial \psi}{\partial t} = H\psi$$

Or $i\hbar \frac{\partial \psi}{\partial t} = E\psi$

Doing product differentiation

$$i\hbar \frac{1}{dt}(\psi e^\theta e^{i\gamma n}) = E\psi$$

$$i\hbar \left(i\frac{\theta}{dt}\Psi + i\frac{\gamma}{dt}\psi + \frac{\partial \psi}{\partial t} e^{i(\theta+\gamma)}\right) = E\psi$$

Now $\theta = -\frac{i}{\hbar}\int E dt$

Differentiating it yields $E\psi$ which gets cancelled with similar term on the right side.

Cleaning up, using Dirac notation around ψ, we get

$$\frac{\partial \gamma}{\partial t}|\psi\rangle = i\frac{\partial |\psi\rangle}{dt}$$

Taking inner product with $\langle\psi|$ and using $\langle\psi|\psi\rangle=1$

$$\frac{\partial \gamma}{\partial t} = i\langle\psi|\frac{\partial}{\partial t}|\psi\rangle$$

$$\gamma = i\int_0^t \langle\psi|\frac{\partial}{\partial t}|\psi\rangle dt$$

We need to reparametrize the equation. The geometric phase does not care about the time, it is path dependent only. So, instead of time, we use L which is the parameter of the loop that changes the Hamiltonian.

$$\gamma = i\oint_l \langle\psi|\frac{\partial}{\partial l}|\psi\rangle dl$$

This is the formula for geometric phase. It was first described by Berry in 1984.

Is Geometric phase for real?

You bet

Foucault Pendulum

Many of you may have seen a big pendulum handing from a ceiling in science museums. As the earth rotates, the plane of pendulum shifts. This is because the earth rotates underneath it. The pendulum completes a loop in 24 hours and goes through parallel transport and picks up a geometric phase that shifts the plane with respect to the earth.

Polarization of Light.

It is also an example of picking up geometric phase.

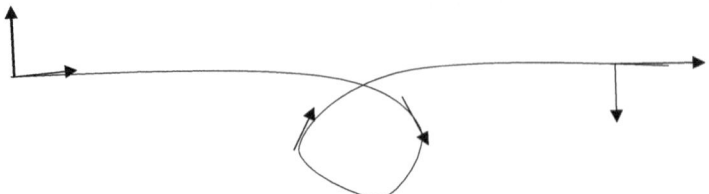

The polarization vector which is perpendicular to the direction of propagation, changes direction as it picks up a geometric phase after going through the loop.

Aharonov-Bohm Effect

This effect again shows the practical effects of geometric phase. To learn about this effect, we need to know some mathematical theorems.

Stokes Theorem

It says that line integral can give us all the information about its surface integral. So, what's going on the boundary tells us what's going on inside.

Is it surprising?

Not really. If someone is crying, we can tell he or she is not happy inside!

Let's describe it mathematically.

$$\oint f.dr \quad \Leftrightarrow \quad \iint \nabla \times f.ds$$

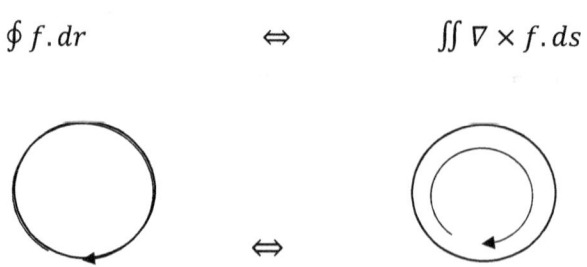

Line Integral   Curl on the inside surface

It is like a boat on the edge of a whirlpool. What is happening to the boat on the edge of the whirlpool (line integral) can tell us how strong the water current is inside the whirlpool (curl of water spinning).

The solenoid is a cylinder with wires wound around it. The electric current in the wires creates magnetic field inside the solenoid.

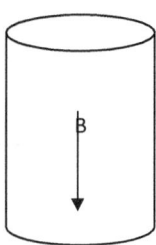

The magnetic field(B) outside the solenoid is zero.

What about magnetic potential(A)?

In classical physics, potential is a theoretical concept, you cannot measure it.

It is the field that is the real thing. It is like saying that I have the potential to lift 25kg bag but only way to measure that would be actually lift the bag.

The problem with the potential is that we can do gauge transformation without affecting the field.

What is a gauge transformation?

Gauge is a measuring device. You want to measure your weight. You can stand on the scale, you get your weight. There of course are other scales that can tell your weight.

Your weight does not depend on the scale or gauge. That is to say weight is gauge invariant. Thus, doing gauge transformation does not change the result.

Do not take this analogy literally. The gauge transformation is a mathematical concept. In Quantum field theory which supersedes quantum mechanics, everything starts with a Lagrangian. It is made of potentials. The Hamiltonian is derived from the Lagrangian. Potentials are fields whose quanta are quantum particles.

How do we do a gauge transformation?

Add gradient of a function to the potential ∇f or $f = \int A dr$

Gradient is ∇f or $f = \int A dr$

Original potential is A_0.

$A = A_0 + \nabla f$

Magnetic field B is given by Maxwell equation $B = \nabla \times A$

$B = \nabla \times (A_0 + \nabla f)$

$\nabla \times \nabla f = 0$

The curl of a gradient is zero, so this gauge transformation of adding gradient of a field to the potential does not contribute to the magnetic field.

Why gradient of a curl is zero?

Gradient means going uphill or downhill. Curl means moving in a circular manner and reaching the starting point like a circle. You cannot go downhill and reach the place where you started!

This is the basis of the Stokes theorem as well. Curls inside the boundary cancel out and do not contribute to the integral.

Going back to the solenoid, the particle can go around it, taking different paths.

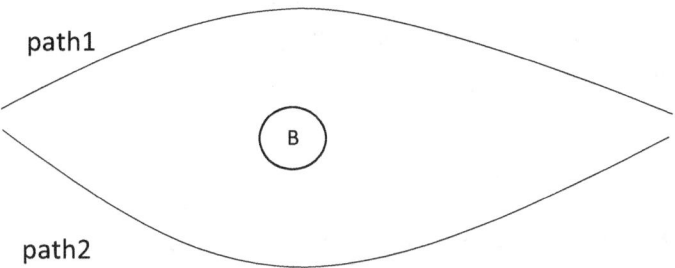

A particle can either take path 1 or 2 around the solenoid to travel across it. The two paths form a loop and we can use Stokes theorem to evaluate it.

$$f = \int A dr \Leftrightarrow \iint \nabla \times A . ds = B \cdot ds = \text{Magnetic flux}$$

Does particle going around the solenoid experience any effects of magnetic field?

The magnetic field outside the solenoid is zero, so it should not be influencing particles outside at all. But as shown above using Stokes theorem, line integral is influenced by the magnetic flux inside.

The answer is the particles do get influenced by the path they take. They pick up a phase which is path dependent. It can be experimentally checked by making two electrons go through the two paths and once they pick up the phase, interference pattern is noted. The waves (electrons) that are in or out pf phase will obviously cause constructive or destructive interference. This is the Aharonov-Bohm effect.

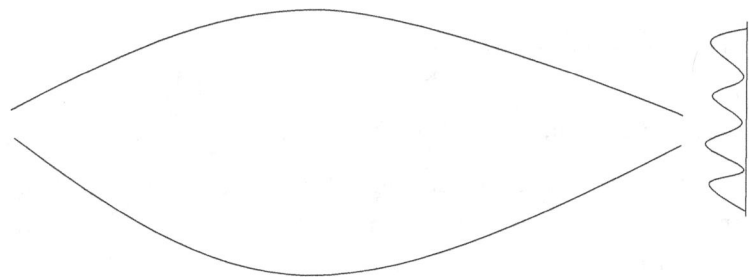

Interference Effect

This effect is surprising. The gauge transformation was supposed to be a mathematical trick as potential can be arbitrarily chosen. But now we see that magnetic potential does produce measurable effect without the presence of the magnetic field.

How do we calculate the phase?

We put the Hamiltonian in the Schrodinger equation and solve it.

The Lagrangian (KE-V) is written in terms of potentials not fields as this is the only way the theory works. The Hamiltonian (KE+V) can be calculated from Lagrangian

$$H = \Sigma qp - L$$

Here q, p are canonical position and momentum.

I won't go into details of Hamiltonian mechanics, you should go back to earlier chapters if you are forgetting the details.

$$\nabla = \frac{\hbar}{i}\nabla - qA$$

$$H = \frac{p^2}{2m} + V = \frac{-\hbar^2}{2m}\nabla^2 + V = \frac{-\hbar^2}{2m}(\frac{\hbar}{i}\nabla - qA)^2 + V$$

Solve the equation with original A. Then do a gauge transformation to A where $A \neq 0$ but B=0 and solve it again and compare.

$$\psi = \psi_0 e^{ig}$$

ψ picks up a phase due to gauge transformation.

Note that $|\psi|^2 \underline{=} |\psi_0|^2$ as e^{ig} cancel out with e^{-ig} when doing complex conjugation to calculate probability. This is true for one electron but if you send two electrons through each path then each will differ by a phase that will not cancel out and that results in interference pattern.

$g = \frac{q}{\hbar}\int A dr \approx \Phi$, magnetic flux.

If we solve Berry phase equation $\gamma = i\oint_l \langle\psi|\frac{\partial}{\partial l}|\psi\rangle dl$

We will find that the geometric phase is nothing but Berry's phase $e^{i\gamma}$.

Scattering

Quantum particles are too small to be observed directly. Collisions or scattering experiments are key to learn about their properties.

A beam of particles is incident upon the target and gets deflected at an angle.

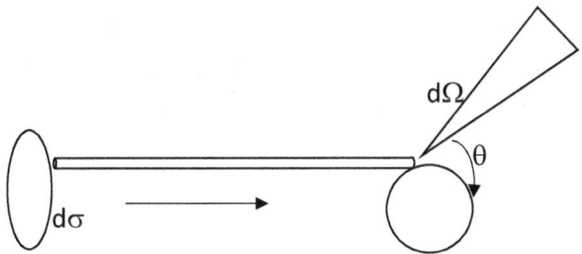

$d\sigma$ is small part of the incident beam

$d\Omega$ is the small part of the deflected beam.

$\frac{d\sigma}{d\Omega}$ is called the differential cross section.

It represents how many incoming particles get deflected per unit solid angle.

It is like a sniper hitting the enemy with bullets. We want to track how many hits a sniper gets by adjusting angle of his shot.

$\sigma = \int \frac{d\sigma}{d\Omega} d\Omega$ is called the total cross section. It is integrating over all the volume.

This is the effective collision zone or war zone.

Let's do a modeling of scattering.

A plane wave hitting a fixed target is a good example.

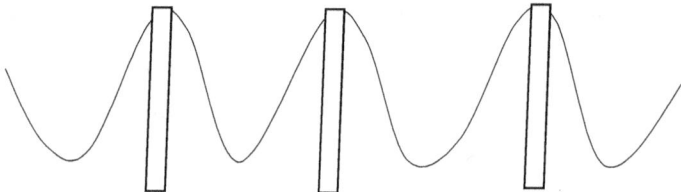

This is a plane wave as wave is not changing with time, so same individual wave packets are coming one after another. The advantage of this approximation is that we can assume constancy of time and deal with time independent Schrodinger equation $H\psi = E\psi$, which is simpler to solve that time-dependent equation.

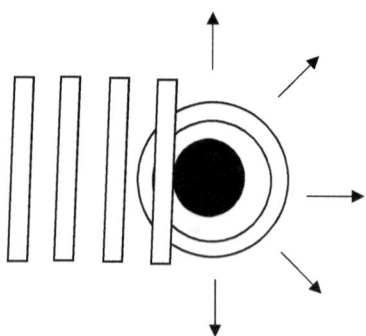

The plane wave hits a fixed target and gets reflected in form of a spherical wave.

Total ψ = incoming wave + outgoing reflected wave

Incoming plane wave= e^{ikx} is a good solution, we have seen it before in finite barrier models.

Outgoing spherical wave= $f(\theta)\dfrac{e^{ikr}}{r}$

1/r makes sense as spherical wave is spreading outwards, so probability decreases with increasing r.

$f(\theta)$ is the scattering amplitude. This is the key that distinguishes how one scattering is different from another. If we throw a rock or a cup of glass at the target, it is the scattering amplitude that will tell us how these two will differ.

The probability of waves is $|\psi|^2$.

If we calculate the probability of incoming and outgoing waves and calculate differential cross section, we will find

$$\dfrac{d\sigma}{d\Omega} = |f(\theta)|^2$$

There are two ways to go about solving the Schrodinger equation.

1. Partial wave Analysis

This method is useful when target is small or potential V is limited in range and the incoming wavelength is long compared to the target.

This makes sense as we want to decompose each incoming wave to see how it gets deflected. A long slow wave length is easier to follow than very rapid short wavelength.

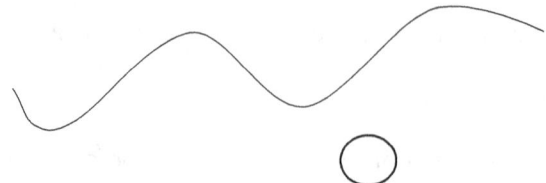

Long wave length and small target

Hψ = Eψ

We are doing to divide ψ into spherical coordinates, similar to what we did to solve the Hydrogen atom.

$$\Psi = R\theta\phi$$

The potential is spherically symmetrical in our approximation. So, it is not varying with angle around z axis (ϕ). This means we don't have to worry about ϕ part of the equation.

This means θ only remains in the angular equation and the result is known to us from hydrogen atom. The solution is Legendre polynomial.

Angular solution, also called Y_L = P cos θ

Where P represent various Legendre polynomials. The polynomials depend on the angular quantum number L.

Recall from hydrogen atom that radial equation was

$$\frac{1}{R}\frac{d}{dr}\left(\frac{r^2 dR}{dr}\right) - \frac{2mr^2}{\hbar^2}(V\text{-}E) = l(l+1)$$

The equation was simplified by change of variable $u = rR$

The radial equation can be written as

$$-\frac{\hbar^2}{2m}\frac{d^2u}{dr^2} + [V + \frac{\hbar^2}{2m}\frac{l(l+1)}{r^2}]U = EU$$

In case you are rusty on the details, go back to the hydrogen atom and review how we derived this equation.

$\frac{l(l+1)}{r^2}$ is the centrifugal term. It wants the wave to get out of the circular motion around the target.

The area around the potential or target can be divided into zones as each zone has slightly different conditions.

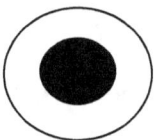

Target is black, around it is an intermediate zone and the rest of the outside is the radiation zone.

In the target zone, potential is non-zero.

In the intermediate zone, potential is zero, but centrifugal term cannot be ignored as there is tendency to push the wave away from circulating around the target.

In the radiation zone, potential is again zero and centrifugal term can be ignored as wave is already far away from the target so the force to keep it away from circulating around the target is negligible.

It is not difficult to see this in daily life. If you are cleaning your car with the garden hose, the pressured water strikes the car, some of water just next to the car will try to circle the car and then it splashes away from the car.

Let's solve the radiation zone first, it is the easiest.

For large r, V=0, centrifugal term= negligible

$$-\frac{\hbar^2}{2m}\frac{d^2U}{dr^2} = EU$$

$\frac{d^2U}{dr^2} = -k^2U$ where k= $\frac{\sqrt{2mE}}{\hbar}$

Solution is well known, we have seen in various examples

U= $Ae^{ikr} + Be^{-ikr}$

Be^{-ikr} term represents an incoming reflected wave, but we are interested in only outgoing reflected wave, so this term can be ignored.

U= Ae^{ikr}

Since $u = rR$, we find

Radial part of the Schrodinger solution is = $\frac{e^{ikr}}{r}$

This is as expected.

In the intermediate zone, there is centrifugal term

$$-\frac{\hbar^2}{2m}\frac{d^2U}{dr^2} + [\frac{\hbar^2}{2m}\frac{l(l+1)}{r^2}]U = EU$$

How do we solve this equation?

We try power series method

$$y = \sum c_n x^n$$

There is singularity in the equation $\frac{1}{r^2}$

If r is zero, then solution blows up.

Here we will use modified power series called Frobenius series.

$y = x^r \sum c_n x^n$, where r is the root.

This is like building your home, there are special bricks that are needed to cover potholes in the house(singularity).

Where do you get these bricks to build the solution to your home?

You go to the nearest Math depot, ask for the bricks.

The math consultant will aptly recognize the solution to the problem as Bessel functions.

$$y = Aj_l + Bn_l$$

j is called Bessel function of the first kind and it is finite at the origin. It is like sine of a function.

N is called Bessel function of second kind and it blows up at the origin. It is like cosine of a function.

It is more convenient to express sin + cos into e^{ikx} by Euler formula

Here we go, more functions

$j_l + in_l = h_l$ Henkel function of first kind, like e^{ikx}

$j_l - in_l = h_{l2}$ Henkel function of second kind, like e^{-ikx}

e^{-ikx} is the incoming reflected wave, so we can ignore this term.

Henkel function of first kind is all we need.

Total solution $\psi = e^{ikz} + \sum(2l+1)c_l h_l Y_l$

where Y is angular solution and 2l +1 is part of factor included in the solution.

We take z direction as the preferred direction of the incoming wave.

Here $z = r\cos\theta$. I know there are too many variable changes to keep track of!

We already deduced that if r is very large, radial solution is like $\dfrac{e^{ikr}}{r}$

Or $h_l = \dfrac{e^{ikr}}{r}$

And solution $\psi = e^{ikz} + f(\theta)\dfrac{e^{ikr}}{r}$

This means $f(\theta) = \sum(2l+1)c_l Y_l$

What we are doing here is each wave with different l will have a different solution. We have to add up all the solutions to know how scattering of each wave takes place.

It looks like a herculean task as l can go from 0 to infinity. Fortunately, with higher l values, waves are much more spread out. Only lower l value waves participate in scattering as higher value waves go unchanged.

This is like an elephant moving and hits a small rock. Only his foot will participate in scattering. The rest of the body pass through the obstacle unscathed.

Incoming wave e^{ikz} can also be converted into Bessel and Hankel functions.

$$e^{ikz} \sim \sum (2l+1) j_l Y_l$$

n_l is not needed in the solution, I won't go into technical details.

Incoming Hankel → Outgoing Hankel function

This leads to a phase shift of the function, $e^{2i\delta e}$

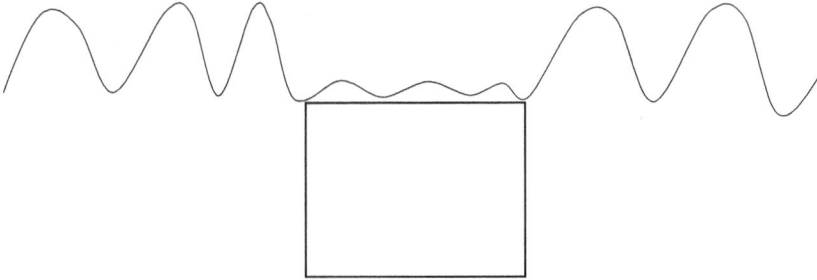

This is similar to if you are walking on a road and hit a pot hole. Your walking rhythm will be disturbed by the pot hole. As you recover, your rhythm will shift as you get scattered by the pot hole and hopefully not fall to the ground.

Hard sphere scattering

When wave hits a hard object, nothing happens to the object, but the wave gets scattered.

At the boundary $\psi = 0$ or

Total wave solution~ $[j_l + c_l h_l] Y_l = 0$

c_l can be calculated.

c_l can be used to then used to calculate $f(\theta)$

Finally, total cross section $\sigma = \int |f|^2 d\Omega$

And if you want to painstakingly do the calculation and integration

$\sigma = 4\pi r^2$

This is 4 times the area of the target.

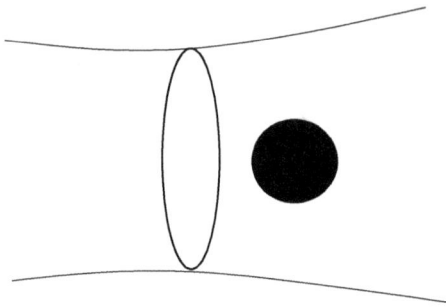

The waves sense the target even before encountering it and start deflecting away.

If there is an accident on the road, it affects people not directly involved in the accident, isn't it?

Born Approximation

There is another way to solve Schrodinger equation besides partial wave analysis.

Let's write Schrodinger equation in a suggestive way

$H\psi = E\psi$

$(KE + PE)\psi = E\psi$

$\dfrac{\hbar^2}{2m}\nabla^2\psi + V\psi = E\psi$

Or $(\nabla^2 + k^2)\psi = Q$

Here k= $\dfrac{\sqrt{2mE}}{\hbar}$ and $Q = \dfrac{2mV}{\hbar^2}\psi$, it is just a clever manipulation.

$(\nabla^2 + k^2)\psi = Q$ is known as Helmholtz equation. It is well known equation in physics and studied well before quantum mechanics was discovered.

To solve it, we need to know about Green function.

$\nabla^2 f(response) = v(source)$ e.g.

$\nabla^2 electric\ potential = charge\ density$

$f = D^{-1}v$

Here we have written the above equation in inverse form. D⁻¹ is inverse operator of ∇^2.

The source is a delta (δ) function. It simply means that one charge is here, another one there and so on. Each charge has a specific location as determined by the delta function.

Where is Green function here?

Let's combine all the individual sources and responses to get the final answer.

Final potential = $\int Gv d^3 r$

G or Green function represents how each source (v) causes a response.

It's part of another well-known equation in physics, called the Poisson equation

$V = \int G\ charge\ density$

V leads to coulomb potential

$$\approx \frac{1}{4\pi\varepsilon_0} \int \frac{\rho}{r-r_1} d^3r$$

$$G = \frac{1}{4\pi\varepsilon_0} \int \frac{1}{r-r_1}$$

Now in the Helmholtz equation

$$(\nabla^2 + k^2)\psi\ response = Q\ source$$

$$\psi = \int GQ dr$$

To this equation, we can always add a constant, ψ_0 is an integration constant.

If $\psi_0 \sim G_0$, so $G_0 + G$ also satisfies this equation.

$$\psi = \psi_0 + \int GQ dr$$

is the integral form of Schrodinger equation.

This equation is not easy to solve as Q also depends on ψ.

In the world of golf, source is the shot played by the player and the response is to get the ball in the hole. Each source is a delta function which means the position of the player in the golf course determines how a shot is played.

Final score $= \int G\ each\ shot$

Writing an equation does not mean it is easy to solve. Even if you know that each hole needs a unique shot or source, it does not mean that it is easy to solve it. Otherwise every golf player will be Tiger Woods.

Likewise, the integral Schrodinger equation is a general equation. Each situation has unique characteristics determined by potential and initial conditions.

How to find Green function?

Obviously if you know initial boundary conditions and how things change (∇) then you can solve the equation

$\nabla G = \delta$

There are two ways to solve the Green function

1. Knowing eigenvalues and eigenfunctions.

$\nabla = L$ is a differential operator.

When this operator acts on ψ, it produces eigenvalues.

$L|\Psi\rangle = \lambda|\Psi\rangle$

G and delta source can be written in terms of their eigenvalues and functions.

$G = \sum g_n \Psi_n$ and $\delta = \sum d_n \Psi_n$

By taking inner product with Ψ_m, all other terms are zero expect when m=n then $\langle \psi_n|\psi_n\rangle$ =1. This trick is frequently used to get rid of sums.

This is how we do paint selection in Home Depot, right! We match the paint of our deck with ones on the shelf at Home Depot and select the one that matches.

So $\lambda g_n = d_n$ Or $g_n = \frac{1}{\lambda_n} d_n$

Thus G= $\sum \frac{1}{\lambda_n} \psi_n^* \psi_n$

This formula works if we know eigenfunctions and values. But again, in our golf analogy it is rewording only to say that each shot is a function that produces an eigenvalue that can be measured, and Green function is a sum of these values.

2. The other method is through the Fourier transform.

$G(r) = \int e^{iqr} g(q) dq$

Here delta source is $\int e^{iqr} dq$

What we have done here is to go from r(position) to q(momentum) space to solve the equation.

It is a different way of looking at the same thing but through a different point of view.

This is like saying that to analyze the golf shots, we go from the real golf course to the computer space, where each shot can be analyzed in slow motion revealing more details to solve the art of scoring holes.

Putting Fourier transform into the Helmholtz equation and not be keeping track of certain factors of π, we get

$(\nabla^2 + k^2)\int e^{iqr} g(q) dq = \int e^{iqr} dq$

$(-q^2 + k^2)g = 1$

Or $g = \frac{1}{k^2 - q^2}$,

putting it back into the green function formula

$G(r) = \int e^{iqr} \frac{1}{k^2 - q^2} d^3q$

We live in 3D world so integration in all 3 directions (d^3q) is needed.

Using spherical coordinates to do the integration

$G(r) = \frac{q^2}{k^2 - q^2} dq \int e^{iqr\cos\theta} \sin\theta d\theta \int d\phi$

ϕ integration is trivial as we assume potential is spherically symmetrical.

Doing θ integration using substitution of variables, after some massaging of formulae, we arrive at

$$G = -\frac{1}{r} \int e^{iqr} \frac{q}{q^2 - k^2} dq$$

This equation has a singularity meaning if k=q then the equation blows up.

Does this equation look easy to you?

Nothing is easy in quantum mechanics, is it?

The answer is complex, literally.

Using complex functions and Cauchy integral formula

$$\oint \frac{f(z)}{z - z_0} dz = 2\pi i f(z_0)$$

Here pole or z_0 is at k. Actually, it is +k or −k. The -k represents incoming scattered wave and thus can be ignored as we are only dealing with outgoing scattered wave. So +k will be used only.

Integral simplifies to $2\pi i e^{ikr}$

See standard texts in the reference section for detailed steps.

$2\pi i$ will get cancelled with other factors of π in the Green function which I have not included.

Finally, Green function is

$$G = -\frac{e^{ikr}}{r}$$

I know you guys must be thinking that I casually introduced complex functions and Cauchy integral formula without giving any intuition behind it. This is what happens in most text books. But I won't let you down! We should take a detour and get some concepts clear.

We have function f that can be integrated to give an answer.

$y = \int f(t)dt$

f could be anything and t can be any variable that it depends on.

If f is very complicated, how do we integrate it?

The trick is to separate the function into real and imaginary parts.

$f = x + iy$

$y = \int f(z)dz = \int xdt + \int iydt$

We have converted f into a complex function.

What's the point?

The integral over a closed surface is very easy if function is well defined.

$\oint_c f(z)dz = 0$

Why is the answer zero?

Let's take simple example first. If a function is independent of the path, then going from one point to another can be chosen freely.

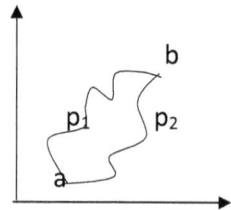

Here path $\int p_1 = -\int p_2$

$\int p_1 + \int p_2 = 0$

The closed loop integral is zero.

Can you think of a function that behaves like this?

Answer is gravitational potential.

It is independent of the path and depends only on the distance between points.

If you go from ground floor to first floor, your gravitation potential rises. It does not matter which path you take, stairs or elevator.

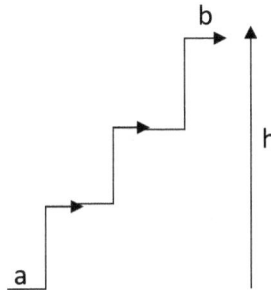

The gravitational potential = mgh, which is obviously independent of the path.

Note that if you take stairs, your shoes will get worn out more as compared to elevator. This means friction is not a conservative force and is thus path dependent.

Since we are dealing with complex functions, we need to learn about complex plane. Let's start with very basic stuff.

Let's imagine you are walking on a street and moving back and forth between point a and b.

As you go from point b to a, real axis is just a straight line, but imaginary axis is half circle from b to a. On the journey back, real axis is straight line from a to b and imaginary axis is half circle from a to b. So, an imaginary circle just represents periodic motion.

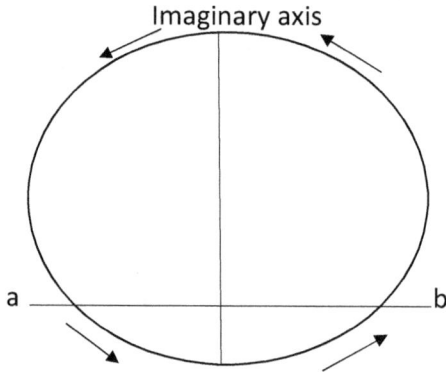

So, when you have a half circle represented like this

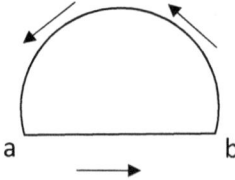

It represents imaginary circle that covers movement from b to a and real axis which covers movement from a to b. Thus, a person came back to same place b, he started from.

No wonder that here $\oint_c f(z)dz = 0$

This is the Cauchy integral theorem.

What if there is a singularity in the function?

Singularity means the function is not well defined at a certain point

$\frac{1}{x-a}$, if x = a then $\frac{1}{0} = \infty$ (not defined)

This is like saying you have a deep sink hole in your street and if you fall into it, what will happen to you is not well defined!

How do we deal with singularity?

In real life, we just go around the sink hole, isn't it?

In complex analysis too, go around the singularity at z_0 as rest of integral is zero. This is called skirting the pole.

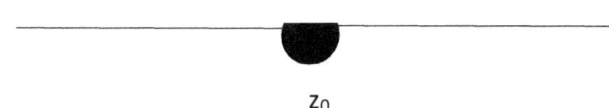
z_0

What to do at z_0?

We can draw small circle around z_0. The circle integral is again zero except at z_0. But we can draw an arbitrarily small circle it is almost localized around z_0.

Integral is thus $\sim 2\pi i$

This is no surprise as circumference of an imaginary circle should be something like this.

Without going into exact derivation, we finally get Cauchy Integral formula

$$\oint \frac{f(z)}{z-z_0} dz = 2\pi i f(z_0)$$

I think this is enough complex analysis for now.

Going back to the Green function and Schrodinger equation, we found that

$$G = -\frac{e^{ikr}}{r}$$

And $\psi = \psi_0 + \int GQ dr$

Putting Green function into the integral equation, we get

$$\psi = \psi_0 - \int \frac{e^{ik|r-r_0|}}{|r-r_0|} V \psi(r_0) d^3 r_0$$

Let's clarify what is r and r_0.

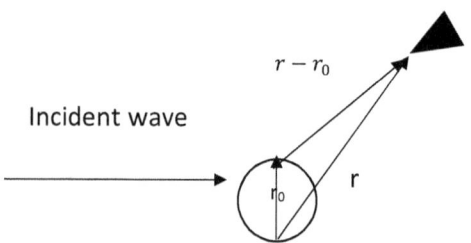

Incident wave

The incident wave is reflected as a spherical wave, and it reaches the detector at either at distance r or r-r_0. Here r_0 is the diameter of the target.

Usually r_0 is much smaller than r as detector is far away from the target. This leads to simplification of the equation.

$$\frac{1}{r-r_0} \sim \frac{1}{r}$$

$k|r - r_0| = k\sqrt{r^2 - 2rr_0 + r_0^2}$

Diving by r, we get a simpler formula $\sim kr - k.\vec{r}_0$

Note here r's are vectors and → hats over them are indicated. When dividing by r, first term becomes scalar so there is no arrow over it. I know it can be confusing and I am not going to track arrows and dot product when writing equations.

So, $\int \frac{e^{ik|r-r_0|}}{|r-r_0|} \approx \frac{e^{ikr}e^{-ikr_0}}{r}$

$\frac{e^{ikr}}{r}$ comes out of the integral since it is constant as we are integrating over r_0.

$\psi = \psi_0 - \frac{e^{ikr}}{r} \int e^{-ikr_0} V \psi(r_0) d^3 r_0$

Scattering amplitude is $\int e^{-ikr_0} V \psi(r_0) d^3 r_0$.

Born approximation

If V is weak then incident wave is not altered much, like hitting a soft target.

We can replace incident wave into the scattering amplitude to simplify the equation.

$\psi = \psi_0 - \frac{e^{ikr}}{r} \int e^{-ikr_0} V \psi_0 d^3 r_0$

Incident wave $\psi_0 = e^{ik_i r}$

Then $\int e^{-ik_f r_0} e^{ik_i r}$ or $\int e^{i(k_f - k_i)r_0}$

Here f and i subscripts represent final (after scattering) and initial (before scattering).

$\psi = \psi_0 - \frac{e^{ikr}}{r} \int e^{i(k_f - k_i)r_0} V d^3 r_0$

You could also write $\int e^{-ikr_0} V \psi_0 d^3 r_0 = \langle \psi_0 | V | \psi \rangle$

$$\psi = \psi_0 - \frac{e^{ikr}}{r} \langle \psi_0 | V | \psi \rangle$$

This is the so called Lippman- Schwinger equation.

These equations can be used to calculate various type of potentials like spherically symmetrical potential. Since we are doing things at conceptual level only, I will skip specific examples. If you dare, you can try solving examples in various textbooks.

Born Series

This should not be confused with Bourne movie series.

Since we assumed incident wave was not altered much and used it in the integral, this process can be repeated over and over to increase accuracy of the results.

Zero order   Incident wave ψ_0 = Scattered wave

First order $\psi_1 = \int GV\psi_0$

Second order $\psi_2 = \int GV\psi_1$

Third order $\psi_3 = \int GV\psi_2$

And so on.

Born series is not too different from a movie series or drama series.

An ordinary guy gets hit after hit from life situations and by end of the series becomes a superhero.

A plane wave is bent again and again with each order to become a scattered wave.

In summary, there are two ways to solve the scattering problem.

Partial wave analysis and Born approximation.

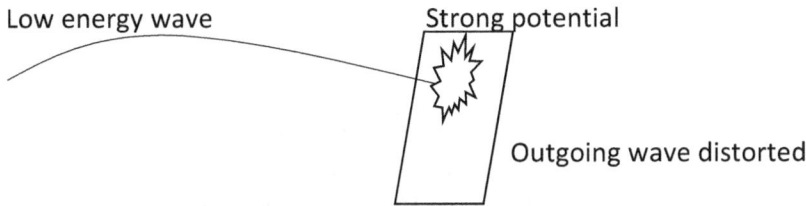

You throw a balloon on a door and it hits the door hard and get splashed making a mess. This is the place to use partial wave analysis.

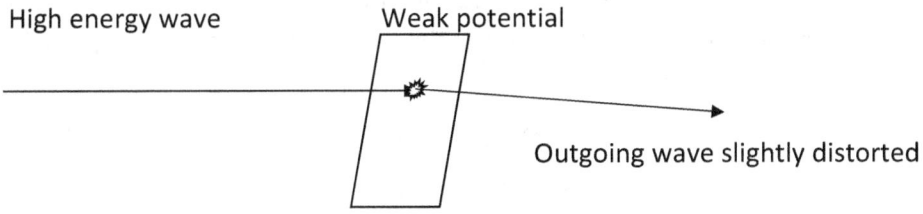

This is like firing a bullet at a window, it will not affect the trajectory much if window is thin. This is the place for Born approximation.

The term $\langle \psi_0 | V | \psi \rangle$ in the scattering amplitude simply implies how incoming bullet (ψ_0) interacts with mirror(V) and gets distorted into outgoing bullet (ψ).

Chapter 10
Standard Model

The Standard model explains the properties of all known elementary particles. It provides classification and order to the elementary particles. It is like periodic table of atoms. The standard model is based on quantum field theory. We will be going into details of quantum field theory in later chapters. The first job is to provide basic facts about the standard model. We need to familiarize with the actors of the Universal soap opera.

At the core, elementary particles are either fermions or bosons. The fermions are ½ integer spin particles that obey Pauli exclusion principle. The fermions do not like each other and cannot occupy same state. The bosons are integer spin particles which can exist in the same state.

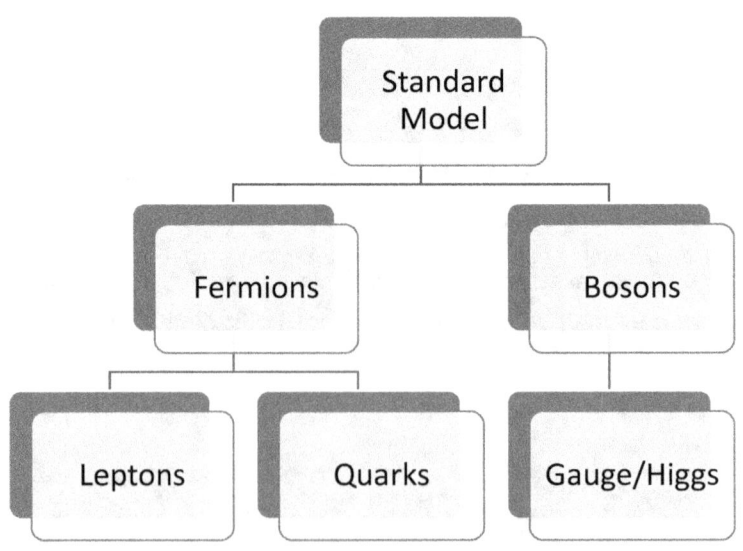

Leptons

Leptons are particles of lighter mass. There are three generations of leptons

1. Electron(e) and its neutrino(v_e)
2. Muon (μ) and its neutrino (v_μ)
3. Tau (τ) and its neutrino (v_τ)

Each of the six leptons have their corresponding anti-particle e.g. anti-electron or positron is the anti-particle of electron, anti-muon is the anti-particle of muon and so on. Each lepton carries lepton number and electric charge number. The numbers are reversed for anti-particles. The exact lepton number and charge for each lepton can be found be found in textbooks and I will not overload you with facts here. The study of properties of electrons and leptons is part of Quantum Electrodynamics.

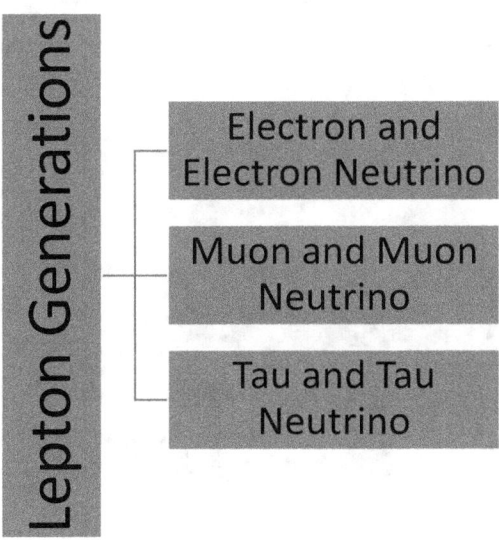

Quarks

The quarks form the nuclei of atoms. They cannot be found in isolation due to confinement hypothesis which we will study in depth later. The quarks have three generations as well.

1. Up and Down quarks
2. Strange and Charm quarks
3. Top and Bottom quarks

The quarks carry color charge. There are three colors-red, blue and green. Colors are just names and have nothing to do with everyday colors. The names of elementary particles should not be taken literally as many of them are silly and not related to underlying mechanisms. The quarks have fractional electric charge e.g. Up quark has + 2/3 charge.

The six quarks have their corresponding anti-quarks as well. The color takes part in strong interactions and is part of Quantum Chromodynamics.

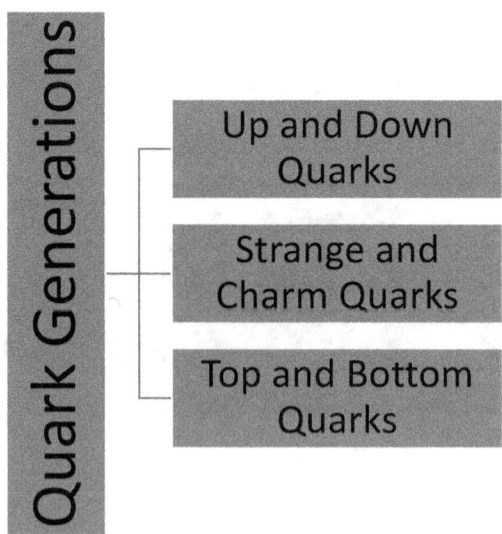

The quarks and leptons are arranged in generations as particles in a generation couple to each other and participate in identical processes. The higher generation is usually heavier than the lower generation.

This is not too different than human generations. Baby boomers share common characteristics that are different than generation Y and millennials.

In case you are wondering what's up with silly names of elementary particles, it is because when those particles were discovered, physicists were skeptical about their existence and it reflects their sense of humor.

The color of quarks is like spin, you could have called it spin too but that name is already taken. After all what's in a name? as Shakespeare so aptly said. The color mathematics is similar to spin, which we will study in detail in the chapter on chromodynamics. The color brings in another variable to fermions so that three quarks with similar spin can coexist in the same state as long as they are of different color, thus obeying Pauli exclusion principle.

The strange quark was discovered when Lambda particle was found to have a longer life span than expected. This was "strange" as strong interactions of known quarks should result in shorter life spans. It was postulated that strange quark gets converted into another quark, which is only possible through weak interaction.

The weak interactions have longer life span. This happens because strangeness is preserved in strong interactions and if lighter quarks do not carry strange quark then the process will not occur in order to preserve strangeness. There is no such preservation in weak interactions.

The point is that each particle has a story behind its discovery and name. It's not very illuminating to tell each story here. It is much more rewarding to know the basic concepts that underlay these processes than go into the history of each particle discovery.

Gauge Bosons

The gauge bosons are mediators of interactions. In quantum field theory all forces are mediated by gauge bosons. They are the messengers of interactions. The force between particles is nothing but gauge bosons interacting with the affected particles. In classical electrodynamics, electrons repel each other via electric fields, there is no role for any mediators. In QFT, the particles are the quanta of the field which means particles come out of the field not the other way around. The electrons repel each other by exchanging photons that are the gauge bosons for electrodynamics.

1. Photons – mediators of electrodynamics or electromagnetic interactions
2. Gluons- mediators of quantum chromodynamics or strong interactions.
3. W and Z bosons are mediators of weak interactions.

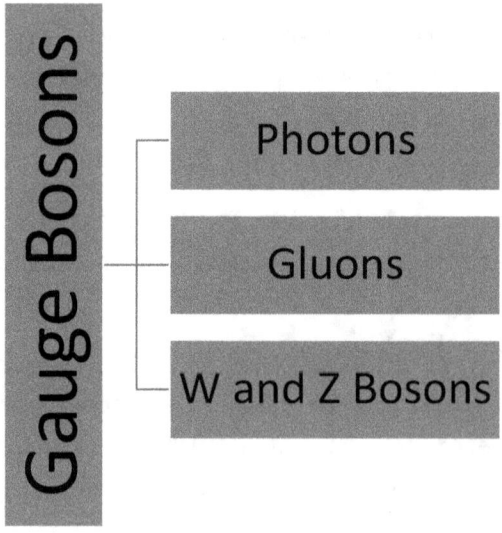

Higgs Boson

It is not a mediator of any interactions. It is the quanta of the Higgs field which is responsible for giving mass to many fundamental particles. Higgs boson is heavy and decays quickly into other particles. Higgs boson was first detected in 2012 at the Large Hadron Collider. It is the latest kid on the block. Why Higgs has a certain mass is a matter of speculation. Various theories like super symmetry have been proposed to explain the mass of Higgs boson.

Composite Particles

Hadrons are composite particles made of quarks, held together by the strong force. They are heavy as compared to lighter leptons. The hadrons are of two types.

1. Baryons- They are made of three quarks e.g. protons and neutrons.

2. Mesons- They are made of quark-antiquark pair e.g. Pion and Kaons.

Units in Particle Physics

A short review of units used in particle physics will come in handy. It is not a difficult topic.

Electron volt is the basic unit used in particle physics.

1 electron volt is the energy gained by the electron as it moves across a potential difference of 1 volt. Since most of the particle physics experiments are done in particle accelerators, this unit makes sense.

1 volt is 1 joule /coulomb,

multiplying by elementary electron charge= 1.6×10^{-19} C

So, 1 eV = 1.6×10^{-19} J

1 MeV= 10^6 eV

1GeV= 10^9 eV

1 TeV = 10^{12} eV

How much energies are these in real life?

Just remember 1 joule of energy is required to lifting a 100g chocolate bar from ground to a height of 1 m. By comparison eV is a very small amount of energy. But we are dealing with individual elementary particles not trillions of particles that are there in a 100g of chocolate bar.

eV can be used as a unit of mass.

$E=mc^2$ so m= eV/c^2

c can be chosen to be 1 unit, called natural unit of light.

e.g. mass of electron is 0.511 MeV, mass of proton is 938.27 MeV.

Particle Accelerators

Modern physics is done in particle accelerators. The properties of elementary particles, discovery of new particles and exotic processes are studied in particle accelerators. High energy collisions are needed to study heavy particles as lot of energy is needed to penetrate what's inside them. The particle accelerators are the only way to generate high energy collisions. Two beams of particles are accelerated and made to collide heads on. This way you get double impact and the energy goes into generating new matter. It is preferred to hitting beam with a fixed target as energy is not transferred efficiently into the collision. Of course, colliding such small elementary particles heads on is a technical challenge in precision. That's why there is huge cost involved in building large accelerators. The most powerful accelerator so far is The Large Hadron Collider (LHC) located in Geneva. It took billions of dollars and decades of preparation to complete.

The charged particles are accelerated by making them move across an electric field. The particles move in bunches, made of billions of protons. The beam made of bunches of particles moves in a tube divided into chambers. You do not want a single tube with electric plates at the end as there is limit to the amount of charge that a plate can take. So instead, chambers with fluctuating electric fields are a convenient way to keep the charges moving in the desired direction. The electric fields are fluctuating as we want to switch from positive to negative charge and vice versa so that charged particle is always accelerated and facing the right kind of charge. The longer the process of acceleration continues, the higher the energy gained by charged particles. The accelerator tube at LHC is 27km long. The energy generated per proton is 6.5 TeV. So, in a head-on collision 13 TeV of energy is available.

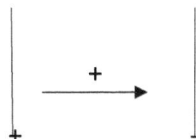

The positive charged particle is accelerated towards the negatively charged plate.

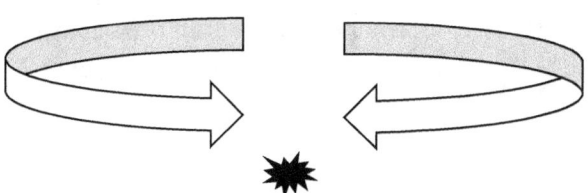

Head-on collision of colliding beams

The beams are directed precisely and not allowed to wander by powerful magnetic fields. There is a generator of charged particles that feeds the accelerator. LHC does proton-proton collisions as protons are complex structures and complicated processes of quark, gluons can occur on colliding them. It increases the probability of discovering new particles. The Higgs boson was one of them that got discovered at the LHC. The electron-positron pair collisions are also done as they help in precision experiments.

The detectors are placed along the accelerator tubes to analyze collisions. They are state of the art instruments and highly complex. There are layers of detection instruments in the detector like seen in a digital camera. ATLAS detector at LHC is one such detector. It has layers of detectors called calorimeters. The detectors absorb particles and detect their path and energy. There is electromagnetic calorimeter, hadron calorimeter and muon calorimeter to absorb various particles and help in detection. All in all, this is a triumph of physics and engineering to create such precise and complex machines. This is only possible with help of collaboration across countries and continents.

Is Standard Model Perfect?

The Standard Model is highly successful in explaining elementary particles and their properties at high precision. However, it is not complete and there are many unanswered questions. The standard model does not predict the mass of particles, they are put in experimentally. There are many other parameters that are also determined experimentally with no theoretical prediction made by the model. The Standard Model predicts that neutrinos have zero mass, but it has been shown experimentally that it is not true. These flaws have prompted physicists to extend the model. There are various theories, but no theory has yet been able to replace the success of the Standard Model.

Chapter 11

Einstein's Special Theory of Relativity

Quantum Mechanics and Special Relativity were brand new theories in the early 20th century. It was natural to join them. It was a remarkable feat of success for theoretical physics that both could be reconciled so elegantly. The credit goes to physicists like Paul Dirac. Special relativity deals with particles moving close to the speed of light. Elementary quantum mechanics has to incorporate the relativity effects that come into picture at higher speeds. Before we learn about relativistic quantum mechanics, we need to know about special relativity.

$E=mc^2$ is part of special relativity. Einstein's reputation was cemented by profound implications of this theory. Unlike quantum mechanics, special relativity is based on very simple principles, but consequences are far reaching.

The core principles of special relativity are

1. Laws of physics are same for all inertial observers.
2. Speed of light is same for all inertial observers.

That's it. With these two principles, whole theory can be built. Isn't it brilliant?

The first principle is easy to understand. The inertial observers mean the observers can have different speed with respect to each other. So, one observer can be standing on the road, another one is going in a car with constant speed. Both are in inertial frame. Whether you are doing experiment in a standing lab or in a lab moving at constant speed, the laws of physics do not change. This makes intuitive sense. Non-inertial frame means acceleration or deceleration is involved. This messes up the laws of physics. If the lab suddenly accelerates, one will feel a pseudo force backwards and the equipment can hit the face! So non-inertial observers do not experience same forces.

Inertial frame of reference= rest or moving at constant speed→ Laws of physics same.

Non- inertial frame of reference = acceleration or deceleration → Laws of physics not same.

The second principle is a curious one. It was not apparent that speed of light should be same for all inertial observers. In fact, it does not make intuitive sense.

The observer in a car moving at 100 miles an hour, shoots a beam of light and measures its speed.

The moving observer will measure speed of light= c.

The observer at rest is also measuring the speed of light at the same time.

So, rest observer should measure speed of light= c + 100 miles an hour

This is because the light was already moving at 100 miles an hour with respect to the rest observer, when light beam started.

The experiments have shown that both land and moving observers measure the same speed of light = c.

Michelson -Morley experiment in 1887 provided the convincing evidence that speed of light is same for all inertial observers. It was designed to check for the presence of ether. It was thought that light needed a medium called ether to travel. So, as earth rotates, the ether moves with respect to the earth. The speed of light should be more when traveling along ether and less when traveling against the direction of ether like if you move in the direction of blowing wind, you have wind at your back!

The experiment split the beam of light in different directions and combined them later by mirrors. It did not find any difference in the speed of light in a given direction.

If speed of light is the same for all observers, then what gives?

$$\text{Speed} = \frac{distance}{time}$$

So, our notion of distance and time has to change to keep the speed of light same.

Consequences of special relativity

1. Violation of simultaneity.

The events that are simultaneous in one frame, may not be the same in another frame.

The classic example is a person traveling on the train. He shines a beam of light that travels towards the front and back of the train compartment at the same time. The person on the train will see both beams of light striking the front and back of the train simultaneously.

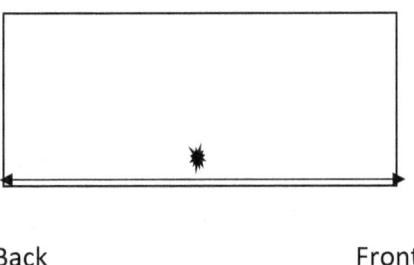

Back   Front

The observer on the train platform who is at rest, will see things differently. He will see the beam of light hitting the back of the train first, followed by the front of the train. This is because back of the train is coming towards the beam of light, so their relative speed is high. The front of the train is moving away from the beam of light, so it gets hit later.

So, the events that are simultaneous for the observer on the train are not simultaneous for the observer on the ground

The events that are connected by light are casually related. When an event happens say an explosion happens at time zero, it spreads outside with maximum limit of speed of light.

Or it spreads like a cone from one time to another.

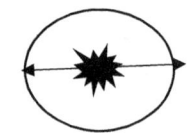

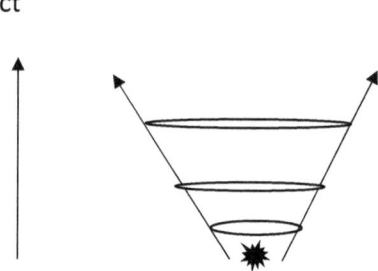

The time flows upwards and is usually combined with speed of light. The cone formed by the flow of events is called the light cone.

The graph of the cone looks like this

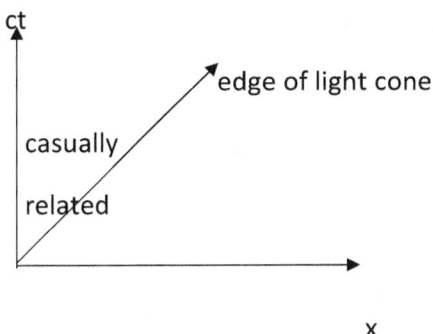

The edge of the cone refers to the angle of 45 degree. On this line the events move at the speed of light. Below it, the distance travelled increases, so speed can exceed the speed of light. The events that take place below this line or outside the light cone are not casually related. The events that place above the light cone line travel at less than the speed of light and are casually related.

The cause and effect are not violated as all observers agree on the order of events

inside the light cone

The events inside the light cone are casually related and called time-like. This is because there is a frame of reference in which events happen at the same place but at different times. The bullet firing from a gun takes place at the same place at different times so these events are casually related and time like.

The events outside the light cone are not causally related and called space like. This is because there is a frame of reference where events happen at different places at the same time. This makes some sense, can someone sitting on a beach see a bullet being fired at one ship and someone die at another ship at the same time? These two events cannot be causally related as bullet firing and reaching a person on another ship has to happen instantly, breaking the speed of light limit which is not allowed.

The moving frames may be tilted but all frames and observers agree on time like events inside the cone.

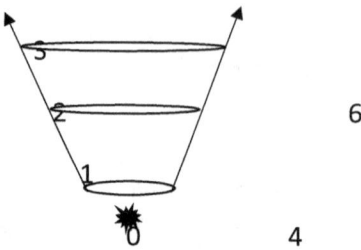

The events 0,1,2 and 3 are causally related as they are inside the light cone.

They could represent

Event 0 = bullet fired

Event 1= hit a person

Event 2= CPR is performed

Event 3 = person dies

Event 4 = NASA probe on the edge of galaxy stops working

Event 6= star explodes

All observers will agree on the events inside the light cone. The person cannot die of bullet injuries and bullet hits him afterwards!

The line that connects the events is called the world line. The space time diagrams shown are called Minkowski diagrams. They represent flat surface as we have not considered curved space time that comes into picture when doing General Theory of Relativity.

The distance between two events(s^2) is frame independent.

$S^2 = c(t_1 - t_2)^2 - (x_1 - x_2)^2$

$S^2 > 0$ is time like, inside the light cone

$S^2 = 0$ is light like, on the light cone line

$S^2 < 0$ is space like, as $x_1 - x_2 > t_1 - t_2$ and speed will exceed speed of light.

Going back to the example of the train, the light reaching the end of the compartment could trigger a reaction. If the light reaching the end of the compartment causes the person at each end to fire a bullet, then the observer on the train will see that the persons on the front and back end of the compartment fired bullets at each other simultaneously. The observer on the ground will see that the person on the back end of the compartment fired first. This is because the events at the end of the train are space like, outside the light cone so observers can disagree on the sequence of events. But the event where torch started and caused the beam of light to travel in each direction is time like and all observers will agree on the fact that torch button was pushed first, and then light beam started, not the other way around. Note that the action of pressing a button and light coming out of the torch happened at the same place but at different times, so it is a time like event. The light reaching ends of the train happens at different places at the same time in one frame, so it is space like event.

Do not worry about guns and bullets in the examples, physicists are peace loving individuals.

2.Time Dilation

It is hard to get your head around that time is relative. We all have a notion of time and our minds have been wired to think that 1 second on earth is 1 second everywhere in the Universe. The notion of absolute time is the basis of Newtonian mechanics.

But in relativity every place has a special clock and different observers can disagree on time.

Let's go to the example of the moving train. Most of the relativity takes place on trains! In physics there is a term for these kinds of experiments. They are called Gedanken or thought experiments.

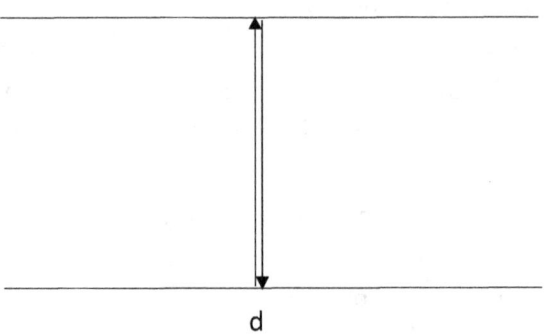
d

A light is shined from the bottom of the train carriage. The beam of light goes up, hits the ceiling and comes back at the same place. The observer on the train will see the beam travel a distance of 2d and take time t to travel that distance

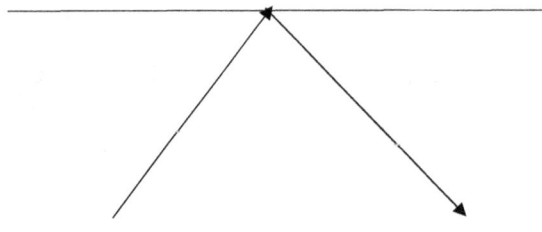

The land-based observer will see a different trajectory of light. It will not be back and forth in a straight line. It will hit the ceiling at an angle and come back at an angle as by the time the light reaches the ceiling, the train has moved by velocity v.

We can draw a triangle and use Pythagoras theorem to calculate the time for land-based observer.

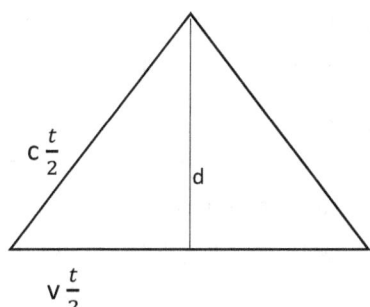

The light covers distance ct and train covers distance vt.

Using Pythagoras theorem, we have

$$\left(\frac{ct}{2}\right)^2 = \left(\frac{vt}{2}\right)^2 + d^2$$

Or

$$T_{land} = \frac{2d}{c\sqrt{1-\frac{v^2}{c^2}}}$$

$$T_{train} = \frac{2d}{c}$$

We define $\gamma = \frac{1}{\sqrt{1-\frac{v^2}{c^2}}}$

$T_{land} = \gamma \, T_{train}$

γ is always more than 1 as v cannot exceed c in the denominator term.

This means time as calculated by the land-based observer, will be longer than the time observed by the observer on the train. The land observer will say that the things are happening slowly on the moving train. This is time dilation. The land-based observer will say "moving clocks run slow".

The peculiar thing is that the person who is in the frame does not notice time is fast or slow, time is running normal, like clockwork!

The train person has no idea that land-based observer is finding that his clock is running slow as when he checks his watch, it is keeping perfect time.

If the train was moving at the speed of light, then the denominator will be zero and time as observed by the land observer will be infinite. The train will appear frozen in time. It is a different matter that massive particles cannot reach speed of light as it requires infinite energy, so only massless particles like photons travel with the speed of light.

Who is moving with respect to whom?

If train is moving with respect to land observer, same can be said by the train observer that the land person is moving, and he is stationary. Now on earth this comparison is not quite valid as train had to accelerate from the rest.

But take the case of an asteroid and earth, observers on both can say that one is stationary, and it is the other who is moving.

Earth observer will say asteroid clock is slow and asteroid observer can say the earth clock is slow. Who is right?

Both are right, time is relative.

It is like an argument, everyone thinks they are right!

Relativity of time can lead to interesting paradoxes.

The twin paradox is the classic one.

If twins get separated, one travels in a space ship to galaxy far far away. The other twin stays on earth. The spaceship is capable of travel near the speed of light. The twin comes back after spaceship journey. It takes 10 years for him, but when he returns to earth, he finds that the earth twin has aged 40 yrs and is an old man.

This is because his clock was slow and earth clock was fast.

But speed is relative, it could be argued that it is the earth that moved away at fast speed from the space ship and so the earth clock should be slow and earth twin should be younger, not older than him.

This is the paradox. The answer is not complicated. The space twin is not in inertial frame all the time. The space ship has to accelerate to reach high speed and then turn back to reach earth. Speed and time are only relative in inertial frames. Non-inertial frames are special frames where other forces come into picture. So, it is the space twin who is younger and the poor earth twin who gets older.

To avoid confusion, the concept of proper time is invented. Proper time is the time observed when at rest with respect to the event. In the train example, proper time is the time observed by the train person as torch was lit on the train.

$T_{land} = \gamma \, T_{proper}$

Some books use τ (tau) to denote proper time.

Time dilation is not a fancy, out of reach concept, it has practical implications.

GPS system is based on corrections to time change based on gravity. Muon has a short life span or proper time. Since muon travels at relativistic speeds.

$T_{earth} = \gamma \, T_{muon}$

The muon life span is increased due to time dilation on the earth. So, muons last long enough to reach the earth lab from the atmosphere.

Sitting in a mathematics class feels like an eternity but playing videogames all day takes no time at all!

3. Length contraction

How do we calculate length?

If we want to calculate distance between earth and a star, we can send a space ship from earth to the star. The space ship will calculate the distance it has traveled, which will be the distance between earth and the star.

Length as calculated by spaceship = v (speed of the space ship) × time taken by the spaceship as per spaceship clock

$L_s = v \times t_s$

The earth-based space station can also calculate the distance

Length as calculated by earth = v × time taken by the spaceship as per earth clock

$L_e = v \times t_e$

Now, $t_e = \gamma\, t_s$

$L_s = v \times \dfrac{t_e}{\gamma}$

Or $L_s = \dfrac{L_e}{\gamma}$

The length measured by moving observer is contracted as compared to length measured by the observer at rest to the measured quantity.

L_e is the proper length as it is measured in the rest frame of the measured quantity.

Both proper time and length are calculated in the rest frame of the measured object. The proper measurements are done at rest with patience, not in a hurry running around.

Length paradox

The length contraction by moving objects can lead to interesting paradoxes.

Imagine there is a fighter pilot with the state-of-the-art aircraft capable of speed close to the speed of light. The pilot sees a small opening in the bunker on the ground in the enemy territory. He has a bomb which is 1 m in length. He calculates the length of the bunker opening but finds its only 0.7m. He is not happy. The bomb will not go in the bunker. The commanding officer on the ground tells him that opening is 1.5 m and he can get the bomb through. The pilot is ordered to release the bomb and he follows through with the order.

What will happen?

Will the bomb go through the bunker or not?

Who is correct?

Both are correct. The length contraction as measured by the pilot has caused the discrepancy in the results.

The command center will see the bomb go through the bunker.

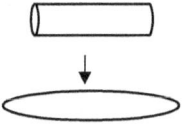

The pilot will also see the bomb go through the bunker, but his calculation is still correct. He will see the bomb slide though the opening.

This again shows the violation of simultaneity. The land observer sees both ends enter the bunker at the same time, but the pilot sees one end going first.

Coordinate Transformation

How quantities change with change of coordinates or frames of reference is the name of the game.

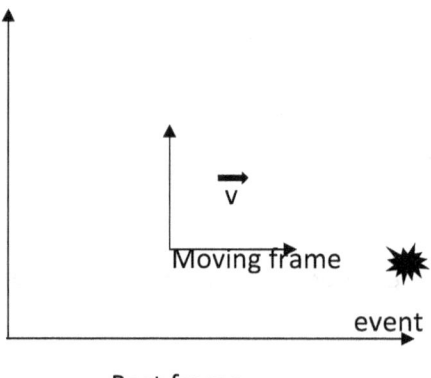

The moving frame has a velocity v. The event could be at rest or moving with velocity u.

The transformation in Newtonian mechanics is pretty straight forward.

If distance of the event from the origin is x in the rest frame, then in the moving frame, distance x' will be

x' = x - vt

The other coordinates remain unchanged as movement is only in the x direction

y'=y

z'=z

Time is universal in Newtonian mechanics

t'=t

These are called Galilean transformation equations.

Lorentz transformations

Transformation equations needs to be modified for special relativity due to time dilation, length contraction and speed of light being constant for all observers.

The equations are called Lorentz transformations.

x' = γ(x − vt)

y'=y

z'=z

$$t' = \gamma\left(t - \frac{v}{c^2}x\right)$$

$$u' = \frac{u_x - v}{1 - \frac{u_x v}{c^2}}$$

If the event is moving with speed of light, say beam of light was started at rest, then the moving frame will observe the same event with speed

$$u' = \frac{c-v}{1 - \frac{cv}{c^2}} \text{ or } \frac{c-v}{1 - \frac{v}{c}} \text{ or } \frac{c\left(1-\frac{v}{c}\right)}{\left(1-\frac{v}{c}\right)} = c$$

This agrees with the basic tenet of special relativity that all observers agree on the speed of light.

Lorentz invariance is the quantity that is unchanged by Lorentz transformation.

It is worth in gold in special relativity.

The space time distance between two events ($s^2 = t^2 - x^2 - y^2 - z$), speed of light and electric charge are some examples of Lorentz invariance.

It is like saying that observers will disagree on who is moving with what speed, but they will agree on the color of their clothes and the smell of their scent.

Relativistic momentum

It is not too hard to figure out.

Classical momentum is $\frac{dp}{dt}$. But which time should we use. Answer is the proper time.

$\frac{dx}{d\tau}, \frac{dy}{d\tau}$ and $\frac{dz}{d\tau}$

But time in a chosen frame is $t = \gamma\tau$

So, $\gamma \frac{dx}{dt} = \gamma u$

Or $p = \gamma m u$

Relativistic Energy

To get to the energy, we first derive the work done to move an object.

$$W = \int F dx = \int \frac{dp}{dt} dx$$

$dx = u dt$, substituting it in the equation

$$\int \frac{dp}{dt} u dt = \int_0^v u \, dp$$

The integral can be solved by doing integration by parts

$$up - \int p \, du$$

Substituting the value of relativistic momentum

$$\frac{mu^2}{\sqrt{1-\frac{u^2}{c^2}}} - \int \frac{mu}{\sqrt{1-\frac{u^2}{c^2}}} du$$

By inspection or guess work, the integral can be solved to

$$\frac{mu^2}{\sqrt{1-\frac{u^2}{c^2}}} + \left| mc^2 \sqrt{1-\frac{u^2}{c^2}} \right|_0^v$$

We can also put integration limits from 0 to v

$$\frac{mv^2}{\sqrt{1-\frac{v^2}{c^2}}} + mc^2 \sqrt{1-\frac{v^2}{c^2}} - mc^2$$

The term $mc^2\sqrt{1-\frac{v^2}{c^2}}$ can be multiplied by $\dfrac{\sqrt{1-\frac{v^2}{c^2}}}{\sqrt{1-\frac{v^2}{c^2}}}$

We will get $\dfrac{mc^2 - mv^2}{\sqrt{1-\frac{v^2}{c^2}}}$

So, we have

$$\dfrac{mv^2}{\sqrt{1-\frac{v^2}{c^2}}} + \dfrac{mc^2}{\sqrt{1-\frac{v^2}{c^2}}} - \dfrac{mv^2}{\sqrt{1-\frac{v^2}{c^2}}} - mc^2$$

Finally, we get the expression

W= $\gamma mc^2 - mc^2$

This work done will get converted into Kinetic Energy of an object as we are taking an object at rest and giving it momentum p.

KE = $\gamma mc^2 - mc^2$

This does not look like ½ mv², the equation that we come to know from high school.

Be patient, first expand γ term in Taylor series expansion

$$\gamma = \dfrac{1}{\sqrt{1-\frac{v^2}{c^2}}} = 1 + \tfrac{1}{2}\dfrac{v^2}{c^2} + \text{ignore higher terms}$$

Taylor series is like any other power series where higher terms are added to increase the accuracy. The exact formula can be found in textbooks.

Basically, taking first few terms of Taylor series is like me drawing Mona Lisa as ☺

We need lot of higher terms to make it look like real Mona Lisa but in Physics, higher terms become less and less important. Basic approximation will do.

Then KE = mc² (1+ ½ $\frac{v^2}{c^2}$) − mc²

Mc² terms cancel out and we are left with ½ mv².

How to interpret the right-hand terms?

γmc^2 is Total Energy= E

mc² is the rest energy.

E = KE + rest energy(mc²)

If momentum is zero, then

E= mc²

The most famous physics equation ever!

Note that if you make speed of light =1, the natural units

Energy and mass are equivalent. They can be converted into one another.

Energy Momentum relation

E = γmc^2

Squaring E² = γ^2 m² c⁴

P= γ mv

Multiplying both sides by c and squaring

$c^2p^2 = c^2\gamma^2 m^2 v^2$

Subtracting it from E^2 term we get

$E^2 - c^2p^2 = \gamma^2 m^2 c^4 - \gamma^2 m^2 v^2 c^2 = \gamma^2 m^2 c^4 (1 - \frac{v^2}{c^2}) = \gamma^2 m^2 c^4 \times \frac{1}{\gamma^2} = m^2 c^4$

Or $E^2 = c^2p^2 + m^2 c^4$

This is called *energy-momentum relation*.

For massless particles

$E = pc$

Note that $E^2 - p^2 = m^2$ is an invariant term. So, all observers agree on the invariant mass.

The invariant mass is independent of the motion of the object. It is the total mass or energy of the object in the rest frame.

In our train analogy, observers will not disagree on the number of people in the train if they rely on the camera that was installed in the train in its rest frame.

Relativistic Notation

Special relativity is full of notation. It can be confusing if you do not know the meaning behind notation.

Four vector

In special relativity, space and time are combined, so there are 4 vector components-one time and three space coordinates.

x^μ is the position four vector where time component is ct and x, y and z are space directions. Usually the space time directions are written as x^0, x^1, x^2 and x^3.

p^μ is the momentum four vector. The components are γmc, γmv_x, γmv_y and γmv_z

To make thing more confusing, they can be written x_μ and p_μ as well.

We need to review tensor notation to get some intuition into it.

v^μ is called contra-variant vector.

v_μ is called co-variant vector.

The difference between them depends upon how these vectors change when coordinates are changed.

The co-variant vector will change along with the change of the coordinate system.

If the coordinate system is made bigger, then co-variant vector will become bigger as well.

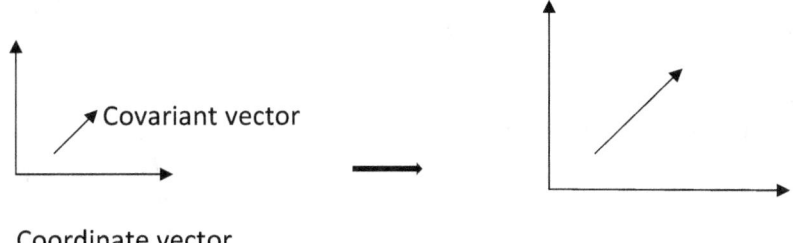

Covariant vector

Coordinate vector

The contra-variant vector goes in the opposite direction to the change of coordinate system. If the coordinate vector gets bigger then contra-variant vector will get smaller. A simple example is going from 1000 m to 1 km. The coordinate vector gets bigger, from m to km but the contra-variant vector gets smaller, from 1000 to 1.

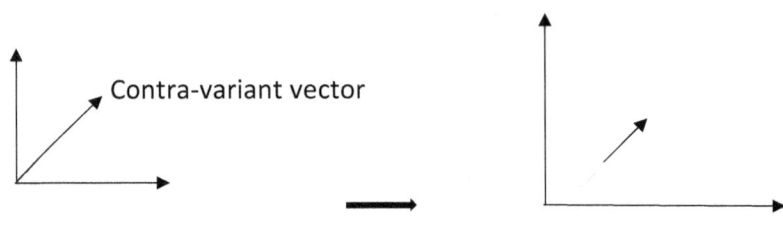

Contra-variant vector

Coordinate vector

The metric tensor can covert covariant vector into contra-variant vector and vice versa.

$v^\mu M_{\mu\nu} = v_\nu$

$v_\mu M^{\mu\nu} = v^\nu$

Note that the upper and lower μ indices contract to convert covariant and contra-variant vectors.

The metric is a tensor and is unique to the conversion. It can be written in a matrix form. In the above example, it is a rank 2 tensor with two indices of matrix elements.

Are you looking for more intuition into metrics and vectors?

Ok, let us take an analogy from finance.

$W_\$$ is the co-variant wealth vector. As you make more $, the wealth vector increases as well.

$W^\$$ is the contra-variant vector. As you add more $ to your debt, the wealth vector decreases. In contra-variant form $ gets a negative sign.

$W_\$ \, B^{\$\$(-)} = W^{\$(-)}$

Here the metric tensor is the balance sheet. It tells us how much money or debt you have, and it can change the direction of your wealth vector.

The metric tensor for special relativity is Minkowski tensor. It represents flat space time. The curved space time due to gravity can get more complicated

$$\eta_{\mu\nu} = \eta^{\mu\nu} = \begin{pmatrix} 1 & 0 & 0 & 0 \\ 0 & -1 & 0 & 0 \\ 0 & 0 & -1 & 0 \\ 0 & 0 & 0 & -1 \end{pmatrix}$$

$X^\mu = \sum \eta^{\mu\nu} X_\nu$, the sum is over all the four space time coordinates (0 to 3)

$X_\mu = \sum \eta_{\mu\nu} X^\nu$

Also, $X^\mu = \begin{pmatrix} ct \\ x \\ y \\ z \end{pmatrix}$ and $X_\mu = \begin{pmatrix} ct \\ -x \\ -y \\ -z \end{pmatrix}$

$X^\mu X_\mu = c^2 t^2 - x^2 - y^2 - z^2$ is the space-time distance called s^2 and is a Lorentz invariant quantity.

$X^\mu X^\mu = c^2 t^2 + x^2 + y^2 + z^2$ is not Lorentz invariant term and is thus not used.

Armed with the basics of special relativity, I would recommend you go through Einstein's relativity papers that are available online. He wrote four papers in 1905 that changed the world of physics forever. 1905 is called the miracle year or Annus mirabilis. You will be able to better appreciate the genius of Einstein rather than relying on pop culture clichés.

Chapter 12

Relativistic Quantum Mechanics

How can we reconcile special relativity with quantum mechanics?

The Schrodinger equation of quantum mechanics is not compatible with special theory of relativity. The speed of light is c for all observers. It has to be built into the equations. The equations need to be Lorentz invariant meaning all observers should agree with them irrespective of their speed.

Let's start with the basic equation of special relativity

$E^2 = p^2 c^2 + m^2 c^4$

Let's put c =1 and eliminate it from the equations for the time being.

$E^2 - p^2 - m^2 = 0$

We need to substitute energy and momentum with quantum mechanical operators

$E = i\hbar \frac{\partial}{\partial t}$ $\qquad p = -i\hbar \frac{\partial}{\partial x}, \frac{\partial}{\partial y}, \frac{\partial}{\partial z}$ or $-i\hbar \nabla$

Again letting $\hbar = 1$, we have

$\left(i\frac{\partial}{\partial t}\right)^2 - (-i\nabla)^2 - m^2 = 0$

The operators need to act on the wave function ψ to get anything out, so let's put that in

$$-\frac{\partial^2 \psi}{\partial t^2} + \nabla^2 \psi - m^2 = 0$$

Using special relativity notation and combining space and time coordinates into one

$$\partial^\mu \partial_\mu = \frac{\partial}{\partial t^2} - \frac{\partial}{\partial x^2} - \frac{\partial}{\partial y^2} - \frac{\partial}{\partial z^2}$$

$$-\partial^\mu \partial_\mu - m^2 \psi = 0$$

This is the famous *Klein – Gordon equation*.

Note there is no spin encoded in this equation, so it describes spin zero particles.

Dirac Equation

$E^2 = p^2 c^2 + m^2 c^4$ does not treat time and space on an equal footing.

$$E = \sqrt{P^2 + m^2}$$

$$\frac{\partial}{\partial t} = \sqrt{\nabla^2 + m^2}$$

The space derivatives are squared as opposed to the time derivative.

The Schrodinger equation is

$$i\hbar \frac{\partial \psi}{\partial t} = H\psi$$

Paul Dirac thought why can't time and space derivatives be treated on an equal footing? Instead of working with $E^2 = p^2 + m^2$, why can't we work with $E = p + m$?

The only way to get $E = p + m$ is to factorize.

$a^2 - b^2 = (a+b)(a-b)$

$p^\mu p_\mu - m^2 = 0$

Here $p^\mu p_\mu = (p^0)^2 - p_x^2 - p_y^2 - p_z^2$, p^0 is the time coordinate.

So, $p^\mu p_\mu - m^2 = (p+m)(p-m)$

we will get $p^2 - m^2$ + cross terms (pm, -pm).

The cross terms don't simply cancel out as p has various space and time derivatives.

Let's multiply p by another set of numbers α and γ.

$(\alpha p + m)(\gamma p - m)$

The α and γ have one time and 3 space components

We have $\alpha^0 p_0 - \alpha^1 p_1 ....$ and $\gamma^0 p_0 - \gamma^1 p_1 ....$

α can be eliminated by setting it equal to γ.

So only γ terms are left.

There are terms involving γ^2. We can set them equal to 1 as well.

$\gamma^2 = 1$.

So, $(\gamma^0)^2$, $(\gamma^1)^2$, $(\gamma^2)^2$ and $(\gamma^3)^2 = 1$.

And then there are cross terms involving γ like γ^{01}, γ^{10} etc.

We want γ^{01}, γ^{10} like terms = zero or $\gamma^{01} = -\gamma^{01}$

This means the terms should anti commute, $\gamma^{01} + \gamma^{01} = 0$

Our goal is to get $p^2 - m^2$ from $(p+m)(p-m)$.

What kind of numbers will meet these weird conditions?

Only Dirac would come up with a genius answer, that γ are matrices.

The matrices are 4 × 4.

$$\gamma^i = \begin{pmatrix} 0 & \sigma^i \\ -\sigma^i & 0 \end{pmatrix} \quad \gamma^0 = \begin{pmatrix} I & 0 \\ 0 & -I \end{pmatrix}$$

Here i runs from 1 to 3 (space coordinates).

σ are our familiar 2 × 2 Pauli spin matrices. The spin is automatically included in the Dirac equation.

$E = (\gamma p + m)$ or $(\gamma p - m)$

We pick one solution $(\gamma p - m)$.

We will change the notation slightly, including γ^0, γ^i into γ^μ There are four γ matrices representing one time and 3 space coordinates.

μ runs from 0 (time) to 3 (space coordinates).

Putting i, $\hbar$ and c back in the equation and since p is made of derivatives, we get

$i\hbar \gamma^\mu \partial_\mu \psi - mc\, \psi = 0$

Where $\partial_\mu \psi = \gamma^0 \frac{\partial}{\partial t}, \gamma^1 \frac{\partial}{\partial x}, \gamma^2 \frac{\partial}{\partial y}, \gamma^3 \frac{\partial}{\partial z}$

This is the *Dirac equation*.

Solutions to Dirac Equation

Since $E = p + m$

The momentum can have positive or negative values e.g. left and right moving particles. This means energy can also be positive or negative.

Negative energy value is a disaster!

If energy can be negative, all particles will prefer lower energy negative values and that's not what we see in the universe.

Dirac thought about this problem and he came with an outrageous answer.

What if all the negative energy values are occupied. The positive energy particles cannot go to negative values as there is no place for them as all negative values are already taken. The occupied negative values form the so-called Dirac Sea.

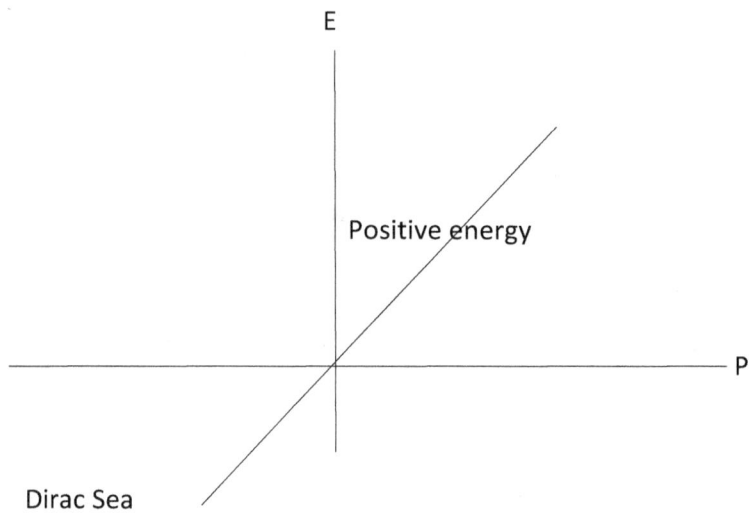

If a negative energy particle is removed from the Dirac Sea, a hole is left. This hole acts like a particle, in fact it is an anti-particle. The physicists were quite skeptical of this theory. But the genius of Dirac was vindicated when the first anti-particle, positron was discovered in 1932. The prediction of entirely new kind of matter, the anti-matter by the Dirac equation is a monumental achievement of theoretical physics, on par with E = mc². It is no surprise that Paul Dirac is considered in such high esteem among physicists.

The modern interpretation of anti-particles come from the quantum field theory. The anti-particles are considered real particles and not holes in the Dirac Sea. The particles and anti-particles are created and annihilated in the vacuum. There is no need for the vacuum to be filled with negative energy particles. Why our universe is made entirely of matter? Why there is so little anti- matter in the universe remains a mystery.

Formal solution to the Dirac equation

The solution to the Schrodinger equation is of the form $\psi = e^{-iET} \psi_{Space}$.

So, let's try something similar

$\psi = e^{-ix \cdot p} \phi$

x.p means $x_\mu p^\mu$ where μ runs from 0 (time) to 3(space coordinates).

$\psi = e^{-iEt+ipx+ipy+ipz} \phi$

ϕ is a spinor and it does not depend on space or time.

Particle at rest

Here space momentum will be zero so only time component is left.

$i\gamma^0 \frac{\partial \psi}{\partial t} - m\psi = 0$ Dirac equation

Putting the value of ψ and differentiating, we get after a bit of cleaning up

$i\gamma^0(-iE)\phi\, e^{-iEt} - m\phi e^{-iEt} = 0$ or

$E\gamma^0 \phi = m\phi$

This is an Eigenvalue equation.

Since γ^0 is 4 × 4, u is a 4-component vector as well.

$$E \begin{pmatrix} 1 & & & \\ & 1 & & \\ & & -1 & \\ & & & -1 \end{pmatrix} u = m \begin{pmatrix} \phi_1 \\ \phi_2 \\ \phi_3 \\ \phi_4 \end{pmatrix}$$

There are 2 positive energy solutions

$E = m\phi$ where ϕ can be $\begin{pmatrix} 1 \\ 0 \\ 0 \\ 0 \end{pmatrix}$ or $\begin{pmatrix} 0 \\ 1 \\ 0 \\ 0 \end{pmatrix}$

The two positive energy solutions refer to spin up $\begin{pmatrix} 1 \\ 0 \end{pmatrix}$ and down $\begin{pmatrix} 0 \\ 1 \end{pmatrix}$

There are 2 negative energy solutions

$E = -m\phi$ where ϕ can be $\begin{pmatrix} 0 \\ 0 \\ 1 \\ 0 \end{pmatrix}$ or $\begin{pmatrix} 0 \\ 0 \\ 0 \\ 1 \end{pmatrix}$

These are the anti-particles with spin up $\begin{pmatrix} 1 \\ 0 \end{pmatrix}$ and down $\begin{pmatrix} 0 \\ 1 \end{pmatrix}$ as well.

Easy Peasy!

Moving particles

Let's put back momentum back in the Dirac equation.

$(\gamma^\mu p_\mu - m)\phi = 0$

$(E\gamma^0 - p_x\gamma^1 - p_y\gamma^2 - p_z\gamma^3 - m)\phi = 0$

Putting value of γ matrices, we get

$$\begin{pmatrix} 1 & 0 \\ 0 & -1 \end{pmatrix} E - \begin{pmatrix} 0 & \sigma^i \\ -\sigma^i & 0 \end{pmatrix}.p - \begin{pmatrix} 1 & 0 \\ 0 & 1 \end{pmatrix} m$$

Matrix of m is trivial, like multiplying by 1.

Combining these matrices by addition results in

$$\begin{pmatrix} E-m & -\sigma^i.p \\ \sigma^i.p & -(E+m) \end{pmatrix} \begin{pmatrix} \phi_A \\ \phi_B \end{pmatrix} = 0$$

Here we converted ϕ into a 2-component vector.

Two equations can be made from the above matrix equation.

$(\sigma^i.p)\phi_B = (E-m)\phi_A$

$(\sigma^i.p)\phi_A = (E+m)\phi_B$

Putting the value of σ matrices, we get

$$\sigma^i \cdot p = \begin{pmatrix} 0 & 1 \\ 1 & 0 \end{pmatrix} p_x + \begin{pmatrix} 0 & -i \\ i & 0 \end{pmatrix} p_y + \begin{pmatrix} 1 & 0 \\ 0 & -1 \end{pmatrix} p_z$$

Adding together

$$\begin{pmatrix} p_z & p_x - ip_y \\ p_x + ip_y & -p_z \end{pmatrix}$$

Let's solve for ϕ_B

$(\sigma^i \cdot p) \phi_A = (E + m) \phi_B$

Or $\phi_B = \dfrac{\sigma \cdot p}{E+m} u_A = \dfrac{1}{E+m} \begin{pmatrix} p_z & p_x - ip_y \\ p_x + ip_y & -p_z \end{pmatrix} \phi_A$

ϕ_A can be $\begin{pmatrix} 1 \\ 0 \end{pmatrix}$ or $\begin{pmatrix} 0 \\ 1 \end{pmatrix}$.

Remember original u had 4 components, so we divided u into 2 component vector ϕ_A and ϕ_B.

$$\phi_B = \begin{pmatrix} p_z & p_x - ip_y \\ p_x + ip_y & -p_z \end{pmatrix} \begin{pmatrix} 1 \\ 0 \end{pmatrix} \text{ or } \begin{pmatrix} p_z & p_x - ip_y \\ p_x + ip_y & -p_z \end{pmatrix} \begin{pmatrix} 0 \\ 1 \end{pmatrix}.$$

$$\phi_B = \begin{pmatrix} p_z \\ p_x + ip_y \end{pmatrix} \text{ or } \begin{pmatrix} p_x - ip_y \\ -p_z \end{pmatrix}$$

Similarly, $\phi_A = \begin{pmatrix} p_x - ip_y \\ -p_z \end{pmatrix}$ or $\begin{pmatrix} p_z \\ p_x + ip_y \end{pmatrix}$

We have now solved the Dirac equation, let's summarize.

$$\phi = \begin{pmatrix} \phi_1 \\ \phi_2 \\ \phi_3 \\ \phi_4 \end{pmatrix} \text{ or } \begin{pmatrix} \phi_A \\ \phi_A \\ \phi_B \\ \phi_B \end{pmatrix}$$

$$\phi_1 = \frac{1}{E+m} \begin{pmatrix} 1 \\ 0 \\ p_z \\ p_x + ip_y \end{pmatrix}, \quad \phi_2 = \frac{1}{E+m} \begin{pmatrix} 0 \\ 1 \\ p_x - ip_y \\ -p_z \end{pmatrix}$$

$$\phi_3 = \frac{1}{E+m} \begin{pmatrix} p_x - ip_y \\ -p_z \\ 0 \\ 1 \end{pmatrix}, \quad \phi_4 = \frac{1}{E+m} \begin{pmatrix} p_z \\ p_x + ip_y \\ 1 \\ 0 \end{pmatrix}$$

We have to put normalization constants and speed of light in the equations for completion.

ϕ_1 and ϕ_2 are positive energy particles with spin up or down.

ϕ_3 and ϕ_4 are negative energy or anti-particles with spin up and down.

That's lot of math but I hope it makes sense to you. We will make use of our politicians to get an intuitive meaning to the math above.

Can we write a Dirac equation for politicians? Let's give it a shot.

$\gamma_p \partial \psi$ = votes for the budget

$$\psi \text{ (US politician)} = \begin{pmatrix} \psi_1 \\ \psi_2 \\ \psi_3 \\ \psi_4 \end{pmatrix}$$

ψ_1 = Moderate Republican

ψ_2 = Conservative Republican

ψ_3 = Moderate Democrat

ψ_4 = Socialist Democrat

γ_p = matrix of how to get votes for the budget proposal.

$$\gamma_p \begin{pmatrix} cut\ tax \\ cut\ tax\ and\ reduce\ spending \\ increase\ tax \\ increase\ tax\ and\ spend\ more \end{pmatrix} \begin{pmatrix} Moderate\ Republican \\ Conservative\ Republican \\ Moderate\ Democrat \\ Socialist\ Democrat \end{pmatrix} \psi = \text{votes}$$

I hope this gives an intuitive meaning to math behind the Dirac equation and γ matrices even though politics is as far apart as possible from the Dirac fermions.

The Dirac fermions are not the only show in town.

There are Majorana fermions. They were postulated by the Japanese scientist Ettore Majorana in 1930's. There is Majorana equation as well.

$$-i \slashed{\partial} \psi + m \psi_c = 0$$

Here $\slashed{\partial}$ is the slash notation, meaning $\gamma \partial$ and it is used to avoid writing γ too often. The spin is summed over as well. We will learn more about the slash notation later on.

The critical point is ψ_c. It is the charge conjugate of ψ.

$$\psi = i \psi^*$$

The ψ only differs from the charge conjugate by difference in charge only.

ψ = positive charge then ψ_c = negative charge.

This implies that Majorana particle is neutral, since it is made of equal positive and negative ψ's.

The neutral Majorana is its own anti-particle!

Does nature have neutral particles that are their own anti-particles?

The leading candidate is neutrino. The debate that neutrino is its own anti-particle is not settled yet.

Weyl Fermions

It is sometimes convenient to distinguish fermions into left and right-handed fermions. This is especially true for weak interactions as left and right-handed particles are treated differently by weak interactions.

$$\psi = \begin{pmatrix} \Psi_L \\ \Psi_R \end{pmatrix}$$

Weyl fermions are massless, so energy and momentum are related by

H = σ . p, σ is spin matrix

σ . p tells us about the helicity, how spin and momentum are related. Is spin vector and momentum vector pointing in the same direction (right handed) or opposite direction (left handed).

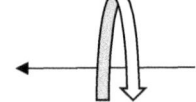

Left handed   Right handed

H = σ . p

Putting quantum operator for H and p

$$i\frac{\partial}{\partial t} = \sigma \cdot \frac{i\partial}{\partial x^\mu}$$

$$i\sigma^0 \frac{\partial}{\partial t} + \sigma \cdot \frac{i\partial}{\partial x^\mu} = 0$$

Using relativistic notation and combining time and space into μ index

$$i\sigma^\mu \partial_\mu \psi = 0$$

is the Weyl equation. ψ is a bi spinor made of left and right handed fermions. We will more about it in the section on weak interactions.

Proca Equation

You thought we are done with the equations? Not quite. The Klein Gordon equation governs spin zero particles, Dirac equation governs spin ½ particles and the Proca equation governs spin 1 particles.

What is a spin 1 particle?

It is the propagator of electro-magnetism, the venerable photon.

Here is the Proca equation for you

$$\partial_\mu F^{\mu\nu} + \left(\frac{mc}{\hbar}\right)^2 A^\nu = 0$$

The electro-magnetism is notational heavy. First, we have to know the notation of classic electro-magnetism, then convert it into relativistic notation.

How to generate a magnetic field B?

By taking curl of the vector magnetic potential A.

$$B = \nabla \times A$$

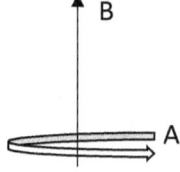

How to generate an electric field?

$$E = -\nabla\phi - \frac{\partial A}{\partial t}$$

Here ϕ is the scalar field. So, potentials lead to fields

In special relativity, we combine space and time

$$A^\mu = \begin{pmatrix}\phi\\A\end{pmatrix} \text{ or } A_\mu = \begin{pmatrix}\phi\\-A\end{pmatrix}$$

How would A change if we want to go from one direction to another?

$\partial_\mu A_\nu - \partial_\nu A_\mu$ where μ and ν are different directions.

This is called the Electro-magnetic tensor, $F_{\mu\nu}$

$F_{\mu\nu} = \partial_\mu A_\nu - \partial_\nu A_\mu$ where μ and ν run from 0(time) to 3(space).

Tensors are used to change one coordinate system to another. $F_{\mu\nu}$ is a rank 2 tensor as it has two indices. The tensors are usually arranged in a matrix form.

$$F_{01} = \frac{\partial(-A_x)}{\partial t} - \frac{\partial \phi}{\partial x} = E_x$$

$$F_{12} = \frac{\partial(-A_y)}{\partial x} - \frac{\partial(A_x)}{\partial y} = -B_z, \text{ this is like curl of A.}$$

And so on.

$F_{\mu\nu}$ makes a 4×4 matrix.

$$\begin{pmatrix} 0 & -E_x & -E_y & -E_z \\ E_x & 0 & -B_z & B_y \\ E_y & B_z & 0 & -B_x \\ E_z & -B_y & B_x & 0 \end{pmatrix}$$

The Electro-magnetic tensor can be used to generate electric and magnetic field in all directions.

Gauge Condition

Physicists loves gauges. Gauges are like paths that lead to the same destination. If all you care about is the destination, then all gauges are equivalent. Gauges are like means to an end, not the end itself!

ϕ = ϕ + any constant is a gauge.

Adding constant does not change anything. The differentiation of a constant is zero, so it does not contribute anything to the equation.

A = A + ∇ f is a gauge as well.

Adding a gradient, ∇ f does not contribute to the magnetic field as curl of a gradient is always zero.

$B = \nabla \times (A + \nabla f)$ but $\nabla \times \nabla f$ is zero.

Gradient means going uphill or downhill. Curl means moving in a circular manner and reaching the starting point like a circle. You cannot go downhill and reach the place where you started!

∇ . A = 0 is called a Coulomb gauge. It means no divergence or spreading out.

A field looks like

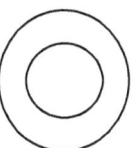

Adding time dimension to the Coulomb leads to the Lorentz gauge.

$$\nabla \cdot A + \frac{\partial \psi}{\partial t} = 0$$

Or in relativistic notation $\partial_\mu A^\mu = 0$

These gauges are used in equations to make calculations simpler.

Maxwell Equations

Maxwell's four equations govern electricity and magnetism. They are crown jewels of theoretical physics. Let's give Maxwell equations relativistic treatment. Maxwell equations can be represented in many ways. We will choose the version which is convenient.

The first two field equations are

$\nabla . E = \rho$

This equation says that the spread of the electric field lines depends on the charge density or how many charges are there. It is like saying that the intensity of light from the Sun depends on how much matter is there in the sun to glow. This equation also tells us that electric monopoles exist, which means positive or negative charges can be isolated unlike magnetic poles.

$$\nabla \times B = J + \frac{\partial E}{\partial t}$$

This equation tells us that a magnetic field can be generated by an electric current and changing electric field.

Here is the relativistic equation that encompasses the above two equations.

$$\partial_\mu F^{\mu\nu} = J^\nu$$

The electromagnetic tensor generates electric and magnetic fields, so a change in the tensor, $\partial_\mu F^{\mu\nu}$ generates the corresponding current J^ν and vice versa.

The remaining two equations are

$$\nabla \cdot B = 0$$

The magnetic monopoles do not exist. The magnetic field lines form a circular loop from north to south poles. They cannot spread out.

N 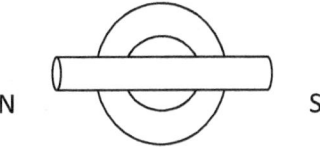 S

$$\nabla \times E = -\frac{\partial B}{\partial t}$$

The final equation is the Faraday law, changing magnetic field generates electric field. This is the basis of electric generators.

The corresponding relativistic equation is

$$\partial_\mu G^{\mu\nu} = 0$$

$G^{\mu\nu}$ is called the dual Electro-magnetic tensor, which is basically $F^{\mu\nu}$ with electric and magnetic fields swapped.

Feynman Diagrams

There is no substitute to visualization. The intuition one gets from seeing something cannot be replaced by mathematical equations. The Feynman diagrams give a pictorial representation to particle interactions. But there is a caveat, don't take diagrams literally. The diagrams don't show exactly how particles interact as it is a complicated process, but they give a line diagram summary of the event.

These are obviously not Feynman diagrams but like them, they show a pictorial representation of a meeting between two people in which a gift is exchanged.

What exactly happened in this meeting is still not clear from the pictures.

There are many possibilities.

There is a possibility that it was a pleasant meeting and may be a birthday gift was exchanged.

There is a possibility that it was just a business delivery.

There is also the possibility that there was a fight and theft took place.

So, you see there are endless scenarios one can come up with.

Similarly, Feynman diagram shows the pictorial representation of particle interactions. But still one has to calculate various scenarios of what exactly happened. Each scenario can be drawn separately. The main diagram is called the tree diagram and various less likely scenarios are called loop diagrams.

What exactly happened = most likely scenario + less likely scenarios

Let's look at some actual Feynman diagrams

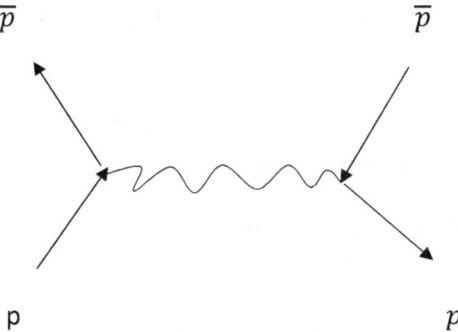

This Feynman diagram shows particle-anti particle collision and annihilation into another particle-antiparticle pair.

$p + \bar{p} \to p + \bar{p}$

Note that arrow of a particle and anti-particle are opposite to each other. This may cause some confusion. The time flows from left to right. On the left side, particle and anti-particles are moving into each other though arrows may not indicate it due to notation, not actual momentum direction.

The wavy line in the middle indicates propagator of the interaction. The propagators pass on the message and move the process forwards.

Here is another important Feynman diagram.

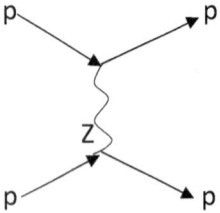

This diagram shows that two protons came close to each other. They got scattered and repelled by the action of the propagator. There is no annihilation in this process.

These diagrams are also called channels. The annihilation channel is called S channel and the above scattering channel is called the T channel.

There is another channel called U channel, it's a bit more complicated so I will just leave it.

I will not go into detail of every Feynman rule. It is not terribly illuminating. We will keep the discussion at the conceptual level.

We need to clarify some terminology used in particle collisions before we do deeper into mathematics of scattering experiments.

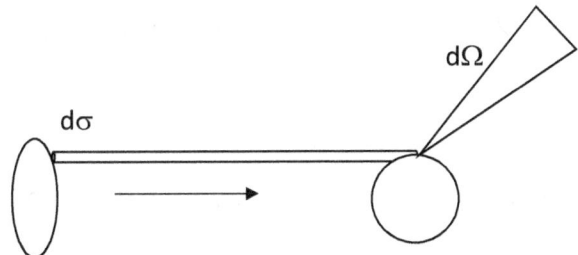

A beam of particles is incident upon the target and gets deflected at an angle.

dσ is small part of the incident beam

dΩ is the small part of the deflected beam.

$\frac{d\sigma}{d\Omega}$ is called the differential cross section.

It represents how many incoming particles get deflected per unit solid angle.

It is like snipers hitting the enemy with bullets. We want to track how many hits a sniper gets by adjusting angle of his shot.

$\sigma = \int \frac{d\sigma}{d\Omega} d\Omega$ is called the total cross section. It is integrating over all the volume.

This is the effective collision zone or war zone.

This is similar to what we learned earlier in the chapter on scattering in elementary quantum mechanics.

We want to calculate scattering cross section to know the strength of the process.

$\sigma \sim \int |M|^2$

Here M is the scattering amplitude of a process.

Each Feynman diagram calculates the scattering amplitude M. The scattering amplitudes of all possible diagrams are then integrated to give the final scattering cross section.

In case of decay of particles, cross section is replaced by decay rate, with the symbol Γ.

$$\Gamma \sim \int |M|^2$$

Like cross section, decay rate is also the final result of integrating various Feynman diagrams of each decay process that is possible in a given situation.

Here is how actual cross section formula looks like

$$\sigma = \frac{1}{E_1 E_2} \int |M|^2 \delta^3(P_1 + P_2 - P_3 - P_4)\delta(E_1 + E_2 - E_3 - E_4) \frac{d^3 P_3}{E_3} \frac{d^3 P_4}{E_4}$$

This looks like monstrosity even after omitting some factors and π's in the equation.

Let's dissect the equation.

Factor $\frac{d^3 P_3}{E_3} \frac{d^3 P_4}{E_4}$ is the phase volume. We have to integrate over the possible values of the phase volume. The particles can move at various speeds and momenta, so all these values form the momentum phase space.

Now $\int \psi\psi dV = 1$, is our usual normalization condition. It means we need to find the particle somewhere. But it is not Lorentz invariant. This is because at higher speed, length contracts. This causes volume to shrink and density to increase. The thing that is conserved is E particles per unit volume.

$\int \psi\psi dV = 2E$ is Lorentz invariant, so all observers regardless of speed agree on this. This is why factors of E are in the denominator in the phase volume.

$\delta^3(P_1 + P_2 - P_3 - P_4)\delta(E_1 + E_2 - E_3 - E_4)$ is a statement of conservation of energy and momenta. If particle 1 and 2 collide and form particle 3 and 4, we expect that initial energy and momenta should match the final energy and momenta. This matching occurs after integrating in all directions.

I will give an investing analogy to give you some intuition behind the equation.

We want to calculate cross section of the stock market. It is the war zone or scattering zone for stocks. Some stocks go up, some down, it's chaos. A well-suited analogy for particle collisions.

$\sigma = \int |\$|^2 \delta$ (money in- money out- commissions) $\times \delta$ (stocks sold – stocks bought) $\times dp^3$ (number of people involved)

This equation kinda make sense, money cannot just be created (except by central banks!). The number of stocks is conserved, and buyers and sellers are conserved. We have to add all transactions to come up with the direction of the stock market.

The key thing here is to calculate scattering amplitude of a stock, $|\$|^2$. This determines why a stock moves higher or lower. Why Apple stock moved up 10 % or Facebook stock tanked 7 %. This is where stock market research matters. The money is made when predictions can be made with regards to stock movements.

The investors study the fundamental factors and macroeconomic events to determine if they should buy or sell a particular stock.

Similarly, in particle physics, scattering amplitude $|M|^2$ determines how likely a particular like $p + \bar{p} \rightarrow p + \bar{p}$ is going to occur. This is our next goal, to learn how to calculate it.

Scattering amplitude

There are certain rules to calculate it.

Incoming particle is called u.

Outgoing particle is called $\bar{u}$.

Incoming anti-particle is called $\bar{v}$.

Outgoing anti-particle is called v.

There is a vertex factor $ig_e \gamma^\mu$, where particle lines meet.

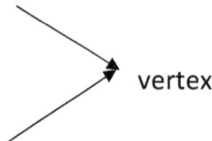
vertex

The propagators have their own factor, $\dfrac{-i\, g_{\mu\nu}}{q^2}$

This is the factor for photon, propagator of Quantum Electrodynamics.

q is the internal momentum of propagators.

The propagators are called virtual particles. This is because virtual particles are not detected in the experiment.

The virtual particles are also called off the shelf particles.

The real particles are on the shelf particles, that obey the equation

$E^2 = p^2 c^2 + m^2 c^4$. The virtual particles do not obey this equation.

If you draw the diagram

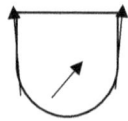

The real particles are on the shelf and the virtual particles are inside it.

M = ∫ (outgoing particle × vertex factor × in coming anti-particle momentum and spin) × propagator factor × (outgoing particle× vertex factor × in coming anti-particle momentum and spin) × δ (energy and momenta conservation)

The actual pairing of particle and anti-particle will depend on each process.

$g_{\mu\nu}$ propagator factor can be absorbed by the vertex factor γ^ν.

$g_{\mu\nu} \gamma^\nu = \gamma_\mu$. This is basic index contraction by multiplying the matrices.

Let's look at the following process, electron-positron scattering

$e^+ e^- \rightarrow \mu^+ \mu^-$

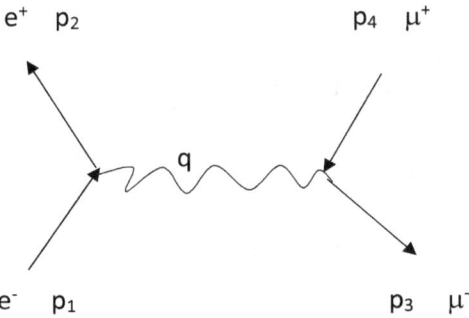

Let's focus on the particle part of the equation

$[\bar{u} \, p_3 \, \gamma^\mu \, v \, p_4] \, [\bar{v} \, p_2 \, \gamma_\mu \, u \, p_1]$

This is M, but we want to calculate M^2, so we need to calculate M^* as well.

To get the complex conjugate, just replace $\bar{u}$ with u, v with $\bar{v}$ etc.

$M^* = [\bar{v} \, p_4 \, \gamma_\nu \, u \, p_3] \, [\bar{u} \, p_1 \, \gamma^\nu \, v \, p_2]$

I haven't mentioned but each particle carries spin as well. But we are not interested in the spin of particles, so we want to get rid of the spin. This is done by averaging over all the initial spins and adding all final spins.

Collect p_3 and p_4 terms first and summing over the spins

$$\sum [\bar{u}v][\overline{uv}]$$

Rearranging

$$\sum [\bar{u}u][\bar{v}v]$$

There is a completeness relation, a particle along with its complex conjugate with all the spins included should obey the Dirac equation.

$$\sum [\bar{u}u] = \gamma\, p + mc, \text{ summing over spin, s= 1,2.}$$

Taking m = 0, an approximation

$$\sum [\bar{u}u] = \gamma\, p \text{ or } \slashed{p} \text{ (Feynman slash notation to avoid writing } \gamma\text{'s)}$$

Similarly, $\sum [\bar{v}v] = \gamma\, p$ or $\slashed{p}$

Finally, $\sum [\bar{u}u][\bar{v}v] = -$ Trace $(\slashed{p}_4\, \gamma\, \gamma\, \slashed{p}_3)$, where I have added back vertex γ as well

There are just too many γ's! I will not even begin to write all γ indices.

The other piece from p_1 and p_2, also leads to a trace after summing over the spins.

$-$Trace $(\slashed{p}_1\, \gamma\, \gamma\, \slashed{p}_2)$

Multiplying both pieces together leads to an extravaganza of multiplication of γ matrices, index contraction. There is a couple of pages of theorems to eliminate γ's and indices. So, in the end we are only left with momentum.

$$|M|^2 = \frac{1}{4} \frac{g^4}{(P+P_2)^4}[\,(p_1 \cdot p_3)(p_2 \cdot p_4) + (p_1 \cdot p_4)(p_2 \cdot p_3)\,]$$

In case you are lost in gamma matrices, think of them as representing direction of motion of particles. We have 4 gamma matrices for each space and time direction. ($\gamma\ \gamma\ \gamma\ \gamma$) multiplication of these matrices leads to a complicated relationship between momenta in various directions. Trace is adding diagonal elements of the products of these matrices. The diagonal elements represent probability of momentum in a certain direction. By taking trace, we are adding probabilities of momentum in various directions and in the end, we are left with product of momentum values only.

$$\begin{pmatrix} p & & & \\ & p & & \\ & & p & \\ & & & p \end{pmatrix}$$ trace= adding diagonal elements of a matrix = adding probabilities of momentum in various directions.

We can simplify complicated product like $[\,(p_1 \cdot p_3)(p_2 \cdot p_4) + (p_1 \cdot p_4)(p_2 \cdot p_3)\,]$ further.

This is done by choosing a specific frame of reference and putting the actual values of momentum. The Center of Mass (COM) reference frame is a convenient choice.

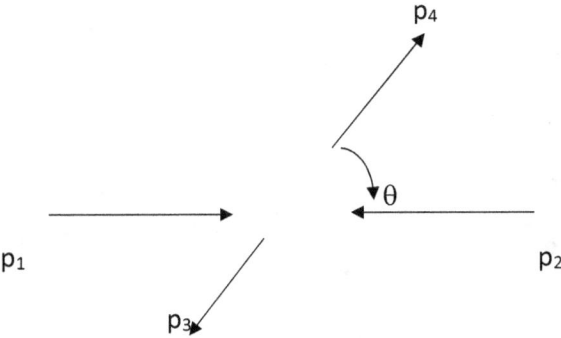

In COM frame, $p_1 = -p_2$

$$p_3 = -p_4$$

If we ignore difference in mass, all have the same energy E

So, $p_1 \cdot p_3 = p_2 \cdot p_4 = E^2 \cos\theta$

The product will simplify to

$|M|^2 = g^4 (1 + \cos^2\theta)$

Once we have M^2, scattering cross section can be calculated.

$\sigma \sim \int |M|^2$

By doing integration over the solid angle, we get a simpler looking expression.

$\sigma \sim \dfrac{\alpha^2}{s}$ where $\alpha = \dfrac{g^2}{4\pi}$ and $s = (p_1 + p_2)^2$

s is called Mandelstam variable. The other two variables are t $(p_1 - p_3)^2$ and u $(p_1 - p_4)^2$. α is the coupling constant. These new variables soak up factors and keep equations lean and clean.

Fortunately, various processes in quantum electrodynamics have their own specific formula and you don't have to go into the tedious process of derivation from Feynman diagrams every time.

$e + \mu \rightarrow e + \mu$ Mott scattering

$e^- + e^- \rightarrow e^- + e^-$ Moller scattering

$e^- + e^+ \rightarrow e^- + e^+$ Bhabha scattering

$\gamma + e^- \rightarrow \gamma + e^-$ Compton scattering

Each process has a formula for scattering amplitude per channel (s, t, and u)

e.g. in electron -positron scattering process we discussed above

$$|M|^2 = \frac{g^4}{s^2}[(p_1 \cdot p_3)(p_2 \cdot p_4) + (p_1 \cdot p_4)(p_2 \cdot p_3)] \sim \frac{t^2+u^2}{s^2}$$

This is because $s = (p_1 + p_2)^2 \approx 2\, p_1 \cdot p_2$.

$T = (p_1 - p_3)^2 \approx -2\, p_1 \cdot p_3$

And so on.

There is a table of scattering amplitudes in terms of Mandelstam variables for each QED process per channel (s, t and u) in various books. So, you just look up these formulae based on each process.

Form factor

How do we see an object?

We shine light on it. To see more clearly, the light has to be bright.

So, we need to hit the particle with high energy to see it more clearly.

The nuclei are made of protons and neutrons. We need to study electron-proton scattering to know the inner structure of protons.

Long wavelength, low resolution.   Short wavelength, high resolution

Differential cross section at low energy, is given by Rutherford formula.

The proton is point like in this approximation. Here electron and proton are moving less than the speed of light, so they are treated non relativistically.

$$\frac{d\sigma}{d\Omega} \sim \frac{\alpha^2}{E^2 \sin^4 \theta/2}$$

At higher energy, the electron is relativistic, so it is moving with lot more energy.

This will give a jolt to the proton, so recoil factor has to be added.

This gives the formula for Mott scattering.

At even higher energy, the proton cannot be treated as a point like structure. It has charge distribution as quarks carry charges. The spin directions are also distributed based on quark orientation inside the proton.

So, we add a form factor. It is made of G_E, charge distribution and G_M, magnetic moment distribution. This form factor is added to the vertex factor and used to calculate the cross section. This gives us the Rosenthal formula.

$$\frac{d\sigma}{d\Omega} \sim \frac{\alpha^2}{E^2 \sin^4 \theta/2} \quad \text{Recoil factor} \times \text{Form factor}$$

Knowing the exact formula is not that important but the concept is. This formula makes intuitive sense. An object that has structure, has to be taken into account during a collision. In boxing, the boxer knows that a punch on face is far more effective than on the arm!

Note this is elastic scattering, nothing is breaking into pieces. The inelastic scattering is much more complicated.

Renormalization

Quantum electrodynamics despite being a very successful theory carries an embarrassing secret. The important calculations in QED result in infinities. The infinities are ignored and soaked up by redefinition of constants and masses. This process is called renormalization.

Let's look at the source of the problem.

The photons are the propagators of QED. This is the main process or tree level diagram. But we have to take other possibilities into account. The other possibility or loop diagrams may include an electron-positron pair that gets created and annihilated.

The propagator factor for photon is straight forward, $\frac{g_e}{q^2}$.

The propagator for electron-proton is a bit complicated. We use the same technique of summing over the spin and calculating the trace of the momenta, that we did in calculating electron-muon collision. The photon momentum is q, so the loop has p, q-p as momenta of the pair. This makes sure that in the end only q is left.

Loop propagator = $\int d^4p \; \frac{Tr[\gamma(\not{p}+mc)\gamma(\not{p}-\not{q}+mc)]}{(\not{p}^2-m^2c^2)[(\not{p}-q)^2-m^2c^2]}$

The mass is non-zero unlike photon, so we use $E = p^2 - m^2c^2$.

Let's count the p's.

$\frac{p^2}{p^4} d^4p$. The d^4p gives $d^3p \, dp$. This trick is used in spherical coordinates as well where $r^3 = r^2 \sin\theta \, d\theta \, d\phi$.

$\int \frac{p^5}{p^4}$ or $\int_p^\infty p \, dp = [\infty - p] = \infty$

This is a disaster. The mathematical term for it is divergence. The technical term is ultraviolet divergence as it happens at high energy or frequency limit.

Propagator = ∞

Then scattering amplitude and cross section will be infinite as well, which is clearly non-sense.

To resolve this mess, we need to express the integral in a different form using various integral mathematics theorems. The final result looks like this

$Loop = g_e^2 \int \left(\frac{dz}{z} - f(x)\right)$

$\int \frac{dz}{z}$ is still divergent. $\int_z^\infty \frac{dz}{z} = \ln z$ or $\ln \infty = \infty$

The integral can be expressed in terms of mass as well as $m^2 = E^2 - p^2$ or $\partial_\mu p$. This is because high energy and momentum implies large mass. We also introduce a cut off, instead of ∞, the upper limit is M. The upper limit is arbitrary, you can choose a high limit as desired.

$\int_{m^2}^{M^2} \frac{dz}{z} = \ln \frac{M^2}{m^2}$, it is not a divergent term.

The process of introducing a cut off to the upper limit of the integral is called regularization. See standard particle physics texts like Griffiths, as mentioned in the reference section for detailed derivation.

The propagator factor = tree level – loop, is like a series expansion

Or $1 - g^2_e [\ln \frac{M^2}{m^2} - f(x)]$

The critical step is to soak $1 - g^2_e \ln \frac{M^2}{m^2}$ term into a new variable g_r.

$g_r = g_e \sqrt{1 - g^2_e \ln \frac{M^2}{m^2}}$

So, propagator factor = $1 + g^2_R f(x)$

The function $f(x)$ is finite and it is of the form $\frac{-q^2}{m^2 c^2}$.

It is finite in a range where momentum (q^2) varies from small to relativistic limit.

This term can be absorbed into g_r as well by stating that total propagator is momentum dependent.

$g_R(q^2) = g_{R(0)} \sqrt{1 + g^2_{R(0)} f\left(\frac{-q^2}{m^2 c^2}\right)}$

The result of all this math trickery is that divergent term, infinities and g_e are gone.

We are left with g_R only.

What should we make of g_r and g_e?

What we thought was the electric charge (coupling constant g_e) is called the bare charge. This means it is only theoretical. The practical charge we measure in the lab depends on the coupling constant g_r. It is momentum dependent and changes based on the energy or momentum level, so it is called running coupling constant. Actually, g_r is the vertex factor and is related to the coupling constant α by $\alpha \sim g_r^2$. But I have neglected to always keep this distinction.

The electron-positron pair in loop diagrams causes screening of the electric charge. The screening is like a cloud, that blocks the sunlight from reaching us. So, when electric charge is measured, screening can cause the electric charge to appear less strong.

All this mathematical trickery does look wishy-washy. If you think like that, you are not alone. Renormalization was controversial from the beginning. Paul Dirac described it like hiding infinites under the rug. You hide small things under the rug, not infinities! Richard Feynman was not convinced either. He termed the procedure as hocus pocus. But there are many defendants of the procedure as well. Steve Weinberg does not see a problem with it, in fact it is essential to explain the lab findings. The present-day physicists have gotten used to renormalization as an essential step in reconciling theory with reality as there is no alternate explanation.

The elementary quantum mechanics involves normalization condition, where we mandate that probability of finding a particle to be 1. The particle has to be somewhere. The renormalization is reconciling the experimental results which are finite and definite with the theory which throws up infinities.

Let's do another boxing analogy. The boxer has all kinds of thoughts in his mind when planning his next move. He could hit this way or that way. These thoughts can be theoretically never ending. But in the end, he does throw a punch. So, you have to put a limit on these thoughts for practical purposes $\int^{\text{limited time}}$. If he takes too long, the other player will make his move and knock him out.

Here's another way to think about renormalization

What actually happens = $\int^{\infty}$ (what can happen-unlimited possibilities) + whatever happens, it happens in a certain limit.

We absorb unlimited possibilities into what actually happens.

So, what actually happens = $\int$ realistic possibility + in a limit

Limit decides where experiment results are valid. It is like saying, stay within the QED limit!

The renormalization procedure will fail at extreme high energy conditions like at the origin of the universe, so a new theory will be needed to explain those conditions.

Electrodynamics of Quarks

Before quarks were discovered, there were several attempts made to unify protons and neutrons. The protons and neutrons are very similar to each other.

Their masses are very similar. The electric charge is the main difference. They can be grouped like spin states

$$p \begin{pmatrix}1\\0\end{pmatrix} \quad n \begin{pmatrix}0\\1\end{pmatrix}$$

Since we know protons are made of quarks, we use the same notation to describe them. The quarks come in many flavors. Up and Down flavored quarks fit perfectly in this notation. It is called iso-spin. It has nothing to do with real spin of the particles, but it is a convenient notation to describe the similarity of quarks.

$$u \begin{pmatrix}1\\0\end{pmatrix} \quad d \begin{pmatrix}0\\1\end{pmatrix}$$

We will use the same math as electron spin. So, there will be total spin and z component of spin. The total spin is denoted by I and z component by I_3.

I_3 can have two values u= +½ (↑), d = -½ (↓)

We combine the above states to get the total spin.

Total spin can be 1 or zero.

There are three states with total spin 1(triplet).

↑↑ ($I = 1, I_3 = +1$)

↓↓ ($I = 1, I_3 = -1$)

↑↓ + ↓↑ ($I = 1, I_3 = 0$)

If you are still confused why the last state has total spin 1, refer to the spin chapter as it's exactly the same math. When we apply lowering operator to this state, we get total spin of 1. The z component is zero, but x or y component may not be zero, that's why total spin is 1. Again, this is just notation, not spin in real dimensions.

There is one singlet state.

↑↓ - ↓↑ ($I = 0, I_3 = 0$)

This again can be checked by applying lowering operators. Here all x, y and z components are zero, thus total spin is zero.

Here is more notation

$2 \otimes 2 = 3 \oplus 1$

This equation is a short hand for saying that iso-spin(two) states combination leads to a triplet state and a singlet state.

Now, protons and neutrons carry three quarks. So, we need more iso-spin combinations.

$2 \otimes 2 \otimes 2 = 2 \otimes (3 \otimes 1) = 4 \oplus 2 \oplus 2$

There are four symmetric combinations (with + sign in combos) leading to total spin $= \frac{3}{2}$

Then there are mixed anti symmetric combinations with total spin = ½.

It looks more complicated, but the underlying principle is the same.

Baryon wave function

How do we write a wave function for a proton or neutron?

Baryon ψ = color × spin × flavor × space

Baryons are fermions which means the wave function has to be anti-symmetric so that exchange of baryons would result in a minus sign.

ψ_{space} = symmetric. The ground state has no angular momentum and looks the same in every direction (like a sphere).

ψ_{color} = anti-symmetric. The colorless singlet anti symmetric states are the only one allowed to exist as free particles. More on it in the next chapter.

So, $\psi_{spin} \times \psi_{flavor}$ = symmetric to keep the final ψ anti symmetric.

There are two ways to make $\psi_{spin} \times \psi_{flavor}$ symmetric

1. Combine the symmetric flavor (total iso-spin $\frac{3}{2}$) and spin states (total spin $\frac{3}{2}$)

This would mean combinations like ↑↓↑ + ↑↑↓ with + signs combined together.

2. Combine mixed flavor and spin states

$\psi_{12} + \psi_{23} + \psi_{13}$

ψ_{12} = (↑↓↑ - ↓↑↑)(udu - duu), there is exchange of 1st and 2nd places.

ψ_{23} = (↑↑↓ - ↑↓↑)(uud - udu), there is exchange of 2nd and 3rd places.

And so on……

What about anti-quarks?

We need to include them in iso-spin symmetry.

quark $\begin{pmatrix} u \\ d \end{pmatrix}$ then anti-quark $\bar{q}$ $\begin{pmatrix} -\bar{d} \\ \bar{u} \end{pmatrix}$

$u \begin{pmatrix} 1 \\ 0 \end{pmatrix}$ $d \begin{pmatrix} 0 \\ 1 \end{pmatrix}$ $\bar{u} \begin{pmatrix} 0 \\ 1 \end{pmatrix}$ $\bar{d} \begin{pmatrix} -1 \\ 0 \end{pmatrix}$

The minus sign ensures that interchange of u to d, $\bar{u}$ to $\bar{d}$ and vice versa does not make any difference.

Combining q $\bar{q}$

(I = 1, I$_3$ = +1) can be formed by combining (I = ½, I$_3$ =+ ½) and (I = ½, I$_3$ =+ ½)

u and - $\bar{d}$ states will do the job.

Similarly, (I = 1, I$_3$ = -1) can be formed by (I = ½, I$_3$ =- ½) and (I = ½, I$_3$ =- ½)

d$\bar{u}$ will do the job.

How do we get I$_3$ = 0?

We need ladder operators to act on above states.

Let's define some ladder operators.

$L_+ u = 0.$ $L_- u = d$ $L_+ \bar{u} = -\bar{d}$ $L_+ d = u$

$L_- d = 0.$ $L_+ \bar{d} = 0$ $L_- \bar{u} = 0.$ $L_- \bar{d} = -u$

$L_- (-u\bar{d})$ will lead to two terms

L_- will first act on u → $-d\bar{d}$, then it acts on $\bar{d}$ → $u\bar{u}$.

So, (I = 1, I$_3$ = 0) state is $u\bar{u} - d\bar{d}$

This is because lowering operator has lowered I$_3$ from 1 to 0.

Similarly, (I = 0, I₃ = 0) can be made by further acting of ladder operators and the result is $u\bar{u} + d\bar{d}$. This is the singlet state. If you try ladder operators on this state, result will be zero.

Here we have $2 \otimes \bar{2} = 3 \oplus 1$, triplet and a singlet state.

What about other quarks?

There is a strange quark, charm quark, top and bottom quarks. Lovely names!

The charm, top and bottom quarks are quite heavy and do not fit well at all with much lighter up and down quarks. Grouping the six quarks together in SU (6) formation does not work. The strange quark is also slightly heavier than up and down quarks. It does not carry iso-spin but it's combination with up and down quarks in SU (3) formation kinda works, not fully.

SU (3)

$$q \begin{pmatrix} u \\ d \\ s \end{pmatrix} \quad \bar{q} \begin{pmatrix} \bar{u} \\ \bar{d} \\ \bar{s} \end{pmatrix}$$

We create more ladder operators to include strange quark

$L_+ s = u \quad L_+ \bar{u} = -\bar{s}$

$L_- u = s \quad L_- \bar{s} = -\bar{u}$

Use these ladder operators to go from higher spin $\frac{3}{2}$ to ½.

$3 \otimes \bar{3} = 8 \oplus 1$

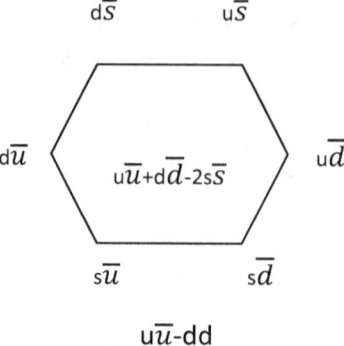

$u\bar{u} + d\bar{d} + s\bar{s}$ singlet

SU (3) is not an exact symmetry so some combinations are not found and are discarded like $u\bar{u} - s\bar{s}$.

The mesons that are $q\bar{q}$ pairs and have strange quarks in them are further classified. The strangeness is considered a quantum number. The strange quark has strangeness of − 1. The other quarks carry 0 strangeness. It is rather arbitrary though.

Mesons can be pseudo scalar or vector mesons.

The pseudo scalar mesons have total spin J=0. The vector mesons have total spin of 1.

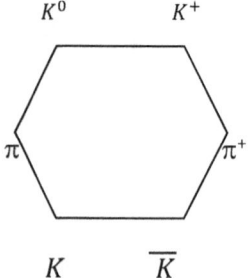

$\pi^0 = u\bar{u} - d\bar{d}$

$\eta \approx u\bar{u} + d\bar{d} - 2s\bar{s}$

$\eta^1 \approx u\bar{u} + d\bar{d} + s\bar{s}$

There is mixing of the above states.

K stands for Kaons; the superscript is for isospin.

π stands for pions.

Then there are sets of vector mesons. I will not state all of them, but these combinations are all particles. It is like counting stars, a never-ending exercise.

You need memory of a zoologist or botanist to remember the names of all particle species!

Chapter 13

Strong & Weak Interactions

QCD is the theory of colors. Don't take the colors literally. It is just a name. We could have labelled them as charge or spin as well. Remember iso-spin is also not real spin (in 3 D space), it's just the name given to properties. The quarks carry three colors- red, green and blue. The anti-quarks carry three anti-colors- anti-red, anti-green and anti- blue. The colors can be represented by a matrix like spin. The matrix is part of SU (3) Group

$$SU(3) = \text{red} \begin{pmatrix} 1 \\ 0 \\ 0 \end{pmatrix}, \text{green} \begin{pmatrix} 0 \\ 1 \\ 0 \end{pmatrix}, \text{blue} \begin{pmatrix} 0 \\ 0 \\ 1 \end{pmatrix}$$

The baryons (protons, neutrons) have three quarks, qqq and thus carry three colors. The mesons are $q\bar{q}$, so they carry one color and anti-color. The interactions involving color are called strong interactions. The mediators of strong interactions are called gluons. The gluons carry one color and anti-color. This is quite different than QED where mediators(photons) carry no charge.

The twist in this color theory is that free particles are colorless. The baryons and mesons are free particles and thus, must be colorless. The gluons and quarks are bound particles. They have never been isolated as they carry net color. So, when combining colors, we have to do it in such a way that colors cancel for free particles.

The colorless invariant state is called singlet state. It is analogous to the singlet spin state where total spin is zero (↑↓ - ↓↑).

Let's first look at the colorless singlet state for baryons.

$$\frac{1}{\sqrt{6}}(rgb - grb + gbr - bgr + brg - rbg)$$

How do we know it is a colorless state?

Well we measure it.

How do we measure color or for that matter anything in quantum mechanics?

We use an operator on the state and get the answer.

The operators used in this scheme are raising and lowering operators, similar to the spin operators. e.g.

C^- is the lowering operator that takes blue to green.

Applying this operator to the singlet state will result in

rgg- grg + ggr − ggr + grg - rgg, which is obviously zero.

The singlet colorless state for mesons is

$$\frac{1}{\sqrt{3}}(r\bar{r} + g\bar{g} + b\bar{b})$$

To see this is also colorless, color operator is again used, but here C^- acts differently on anti-colors. C^- takes red to green and anti-green ($\bar{g}$) to minus red ($-\bar{r}$) resulting in total color zero. This is similar to the math of flavor symmetry operators.

How about gluons?

The gluons cannot exist as free particles. So colorless state is not permitted.

$\frac{1}{\sqrt{3}}(r\bar{r} + g\bar{g} + b\bar{b})$ NOT allowed.

There are eight colored combinations (octet) representing eight types of gluons.

e.g. $(b\bar{r} + r\bar{b})\frac{1}{\sqrt{2}}$, $(r\bar{r} - b\bar{b})\frac{1}{\sqrt{2}}$ and so on....

Note (r$\bar{r}$ - b$\bar{b}$) is colorless but not a singlet state. Confused?

The singlet state means that it is invariant under color transformation which means its colorless irrespective of which colors you use in this combination.

But the above state will not be colorless if you replace the colors with other combinations. So, a singlet state is colorless and invariant to color change.

The colorless state may or may not be a singlet state.

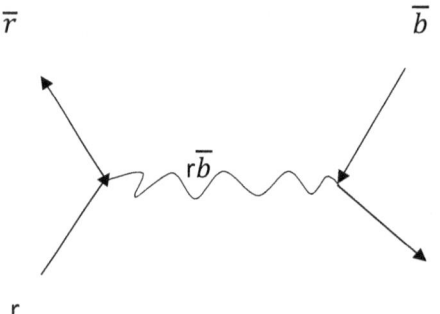

r$\bar{r}$ → b$\bar{b}$ is a typical process of strong interactions, the gluon is the propagator and it carries the color charge as well.

The Pauli and Dirac spin matrices mix up spins and are used to convert one spin state into another. Likewise, we need matrices to mix colors up. The color mixing matrices are called Gell-Mann matrices. There are eight of them as expected as eight gluons are there to act as propagators of mixing. They are labelled from λ^1 to λ^8.

$$\lambda^1 = \begin{pmatrix} 0 & 1 & 0 \\ 1 & 0 & 0 \\ 0 & 0 & 0 \end{pmatrix} \quad \lambda^2 = \begin{pmatrix} 0 & -i & 0 \\ i & 0 & 0 \\ 0 & 0 & 0 \end{pmatrix} \text{ and so on......}$$

The Gell-Mann matrices are used in calculating the color factor which determines if there is attraction or repulsion between colors.

Color factor, $f = \frac{1}{4}\sum \lambda_{ij}\lambda_{kl}$, sum is over all 8 λ matrices. The color goes from i to j and k to l.

e.g. $r\bar{r} \rightarrow b\bar{b}$

Here $r = \begin{pmatrix} 1 \\ 0 \\ 0 \end{pmatrix}$, $b = \begin{pmatrix} 0 \\ 1 \\ 0 \end{pmatrix}$, $\bar{r} = (1\ 0\ 0)$, $\bar{b} = (0\ 1\ 0)$

Multiplying them will result in λ_{12} and λ_{21}. It means that matrix multiplication of $r\bar{b}$ and $\bar{r}b$ will result in 3 × 3 matrix with non-zero values at the following positions-

$$\begin{pmatrix} 0 & x & 0 \\ y & 0 & 0 \\ 0 & 0 & 0 \end{pmatrix}$$

The next step is to look in all 8 λ matrices and find the matrices with non-zero values at the above shown positions. Only λ^1 and λ^2 have non-zero values at those locations. So, we put the values from these λ matrices back into the equation.

$\lambda_{21}^1\ \lambda_{12}^1 + \lambda_{12}^2\ \lambda_{21}^2 = ¼\ [\ (-i \times i) + (1 \times 1)] = ½$

So, the color factor is ½.

This color factor goes into the potential energy formula

$V = -f_{color}\ \alpha_s \frac{\hbar c}{r}$

The negative potential means attraction and positive potential means repulsion. It is same concept used in classical physics if you recall. The negative sign for attraction is a pure convention. Let's take the case of gravitational potential energy.

We define the potential to be zero at infinity. When a particle comes closer to earth, the potential becomes negative as the earth's gravity attracts the particle. We have to do positive work to remove particle from the earth's gravity. Similarly, a positive potential energy between two protons causes repulsion. There is negative work required, which is to say the protons fly away from each other without requiring any work.

Thus, a positive color factor will result in negative potential energy which is attractive and a preferred state. A negative color factor will result in positive potential energy, which is repulsive.

Here are some factors for you.

$r\bar{r} \to r\bar{r} = 1/3$

$r\bar{g} \to r\bar{g} = -1/6$

$r\bar{r} \to g\bar{g} = 1/2$

$rr \to rr = 1/3$

$rg \to rg = -1/6$

and so on.

Let's calculate the final potential for singlet state for a meson.

$\frac{1}{\sqrt{3}} (r\bar{r} + g\bar{g} + b\bar{b})$

To calculate the average potential for the singlet state, we use our familiar expectation value formula $\langle \Psi | V | \Psi \rangle$

$\frac{1}{\sqrt{3}} \times \frac{1}{\sqrt{3}} \langle r\bar{r} + g\bar{g} + b\bar{b} | V | r\bar{r} + g\bar{g} + b\bar{b} \rangle$

There are three $r\bar{r}$ like terms = $\frac{1}{3} \times 3 = 1$

Six cross terms $r\bar{r} \to g\bar{g}$ = ½ × 6 = 3

Color factor = $\frac{1}{3}(1 + 3) = \frac{4}{3}$

$V = -\frac{4}{3}\alpha_s \frac{\hbar c}{r}$

This means attraction, so singlet state is a preferred state.

If we calculate the potential for octet states like $(b\bar{r} + r\bar{b})\frac{1}{\sqrt{2}}$, the potential will be repulsive.

This is reassuring as we expected the singlet colorless state to be preferred as free particles like mesons are supposed to be colorless and the calculation of the potential validates this assertion.

The same game can be played for protons and neutrons (qqq) and there also we will find colorless singlet state has attractive potential.

The potential calculated only works in the short range as long range predictability is poor.

Quark Confinement

Why are quarks confined inside the nucleus? Why can't they be free? Everyone likes freedom! The experiments have failed to detect free quarks so far. The physicists have to come up with some explanation.

The potential energy of the quarks increases with distance.

$V \approx r$

This means that at large distance, it will require infinite work to separate quarks.

Visually, you can think that quarks form flux tubes of energy between them. The lines of flux do not spread out but get more concentrated with increasing distance.

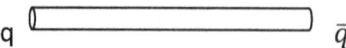
q $\bar{q}$

At large distances, the energy needed to separate them is much more than the energy needed to produce them. So, rather than getting more separated, another quark-anti quark pair is produced. This process keeps repeating and one observes two jets of quark-anti quark pairs in the experiments. The electron positron scattering is a good source to produce quark-anti quark pairs. $e^+ e^- \rightarrow q\bar{q}$

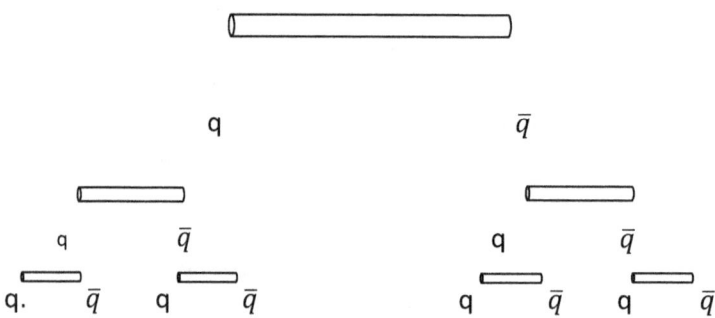

Asymptote freedom

The quarks are involved in strong interactions through gluons, so you expect them to be tightly bound and they are. This is reflected by a strong coupling constant whose value is very big, $\alpha_s \sim 1$. If we want to study quarks at closer and closer distances, we need higher energy probes to hit them. At higher energy, we expect them to be tightly bound in the nucleus as well. But surprisingly, they act as free particles. It means the coupling constant is not constant but it's like a running constant. It changes with energy used.

$$\alpha_s \sim \frac{1}{1+B\ \alpha_S(Q_0^2) \ln\left(\frac{Q^2}{Q_0^2}\right)} \quad \text{where B is } \frac{11N_c - 2N_f}{12\pi}$$

N_c is number of colors, so its value is 3.

N_f is the number of flavors, so its value is 6.

The value of B is greater than 1.

Thus α_s decreases with increased energy or higher momentum(Q).

What's the reason?

If you draw Feynman diagrams, various gluon loops have to be added to calculate α_s. Some of the loops carry negative contributions as well so when totaling all of them, the value decreases due to the negative effect of the gluon loops.

If you look from a distance (low energy), the quark is surrounded by cloud of gluons which keep them tightly bound. As you get very close, you go past the gluon cloud and quarks appear free.

It is like our political leaders. They are surrounded by their entourage and supporters all the time. They cannot speak freely, always toe the party ideology. If we can go past the cloud of supporters, the politicians will appear free and able to say what they think. This happens when they are no longer running for election. That's your asymptote political freedom!

Weak Interactions

The weak interactions involve both leptons and baryons. The propagators of weak interactions are W and Z bosons. W bosons carry charge, W^+ or W^-. The Z boson is neutral. The weak interactions are not the dominant interactions of leptons and baryons. That's why they are called weak. But what makes them weak?

The coupling constant of weak interactions g_w is not weaker than the constants of electrodynamics or QCD. The answer lies in the mass of W and Z bosons. Unlike photons and gluons, W and Z bosons have mass and are heavy. The W and Z bosons are thus sluggish and are not that eager to act as propagators of weak interactions. It is like saying that photons and gluons work like internet. They can instantly relay information across the world. The W and Z bosons are old fashioned postmen, who have to travel on foot to relay messages.

The photons and gluons being massless, carry no polarization along the direction of motion. The polarization vector like electric field is always perpendicular to the direction of motion or momentum. This condition is called Lorentz condition. The coulomb gauge puts ε^0 (time component) to be zero, so only three space directions of polarization are left.

$\varepsilon_\mu p^\mu = 0$. ε is the polarization vector and is always perpendicular to p.

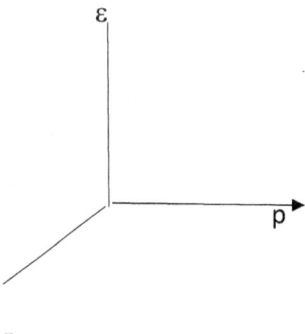

The massive W and Z bosons have the polarization vector in the direction of the motion. So, the Lorentz gauge does not apply here. Why would massive particle get polarization in the direction of motion? The answer is any massive particle can be put to rest by traveling with the same speed or going in its rest frame. If the particle is at rest, then it can be moved in a different direction and this game can be played in all three directions. The massless particle cannot be stopped by going in the rest frame as it moves with speed of light and no one except massless particles move with the speed of light. So, massless particles cannot have polarization vector in its direction of motion.

The hall mark of weak interaction is the distinction made between left and right-handed particles. What are left and right particles in the first place?

Is it the way they write!

No, it is the way they spin. Using the right-hand rule, the fingers represent the spin direction and the thumb represents the spin vector. If the particle is also moving along the spin direction, it is called a right-handed particle. Similarly using the left-handed rule, the particle moving along the thumb of left hand is left handed particle.

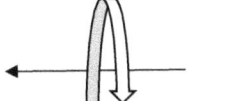
Left handed

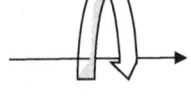
Right handed

The technical name for it is helicity. The right-handed particle can become left handed too. If you outrun the right-handed particle, then in your frame the particle will be moving in the opposite direction, but spin will be unchanged. So, spin and the particle movement will not be the same but reversed, thus making right handed into left handed particle. The way the particle is spinning is independent of its motion and is called chirality. The chirality and helicity are same for massless particles as they move with the speed of light and there is no way to outrun them and change their direction.

Before we describe left and right-handed particles mathematically, we need to know learn about the concept of parity.

Parity is like mirror symmetry. When you stand in front of the mirror, left hand becomes right hand and vice versa.

The hand are vectors as they change in the mirror image. The height is like a scalar, which does not change in the mirror image. When using the right thumb rule, curl of finger is like a vector, thumb is the axial vector as it is up in the mirror image as well.

Parity operator(P) takes x, y and z to -x, -y and -z.

The parity operator has +1 and -1 eigenvalues.

Parity operator × vector = − 1 vector

Parity operator × scalar = + 1 scalar

Parity operator × axial vector = + 1 vector

How to make sense of the parity in terms of fields.

Scalar = $\bar{\psi}\psi$ = (t, x, y and z), no change in signs. You can think of it as probability of finding a particle, there is no directionality to it.

Vector = $\bar{\psi}\gamma^\mu\psi$ = t, -x, -y and -z, change of sign, like curl of fingers

Axial vector = $\bar{\psi}\gamma^5\psi$ = t, x, y and z, no change of sign, like thumb

Left handed parity operator is $(1 - \gamma^5)$. In case you are wondering what is γ^5.

It is nothing but all four gamma matrices multiply together by i.

$$\gamma^5 = i\gamma^0\gamma^1\gamma^2\gamma^3$$

You may wonder how is $(1 - \gamma^5)$ left handed?

The reason is when we are multiplying with the field ψ, γ^μ matrices tag along and it results in $\gamma^\mu - \gamma^\mu\gamma^5$.

γ^μ is vector(finger), $\gamma^\mu\gamma^5$ is (thumb)axial vector. This is like our left-hand thumb, so it represents handedness.

$1 + \gamma^5$ is right handed.

The weak charged interactions involve left handed particles and right-handed anti particles. So, right handed electron is left in the cold while the lefty is given special treatment. The W boson which is a messenger of weak interactions presents message which only left-handed particles can open. Why does nature work this way? Standard model has no answer except saying, that's the way it is!

The weak interactions violate parity. This is the direct result of preference for left handed particles. The physicists for long believed that conservation of parity is the law of nature.

It means that if a process occurs, then it's mirror image should also occur in nature.

The famous example of parity violation is the decay of Cobalt 60.

Co 60 → Ni 60 + electron

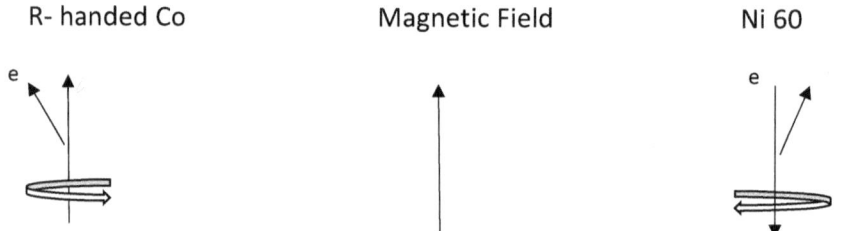

In Co 60, direction of the spin vector, electron and magnetic field is the same.

If mirror symmetry was true, the electron should not show any preference to the spin direction and should be emitted with equal probability in all directions. In particular, you expect that in the mirror process of Ni 60, the electron should be emitted opposite to the direction of the spin as shown above. But this is never seen in experiments. The electron is emitted always in the direction of the spin vector. Thus, parity is violated maximally. The electro-magnetic and strong interactions still preserve parity, but weak interactions violate it.

The parity operator has +1 or -1 eigenvalues. The baryons are given parity of +1 by convention. The anti-particles of baryons have parity of -1. The parity formula, if angular momentum or excited states are included is $(-1)^L$, where L is the angular momentum number. The parity number is multiplicative.

So, a composite particle like meson ($q\bar{q}$) has parity of $+1 \times -1 = -1$

There is charge conjugation operator as well. It changes the sign of the charge, baryon and lepton number but not spin, mass and energy of the particle

Charge conjugation operator × positively charged particle = negatively charged antiparticle

Or $C|P\rangle = |\bar{P}\rangle$

The charge conjugation is violated in weak interactions as well.

The physicists were disappointed by violation of charge conjugation and parity symmetry violations. Physicists love symmetry, so their hope rested on the combined symmetry of charge and parity, together called CP symmetry.

CP operator changes sign of charge and parity.

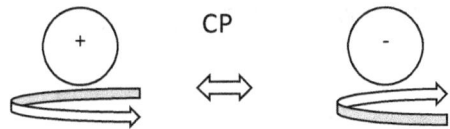

A right handed positively charged particle will become left handed negatively charged antiparticle and vice versa. This is the symmetry between matter and anti-matter. The physicists believed that laws of nature should be the same for particles and antiparticles. Again, nature proved smarter and CP symmetry is violated too in weak interactions. But unlike parity which is maximally violated, the CP violation is small and subtle.

The neutral Kaons violate CP symmetry through decays.

Kaon K = $d\bar{s}$, anti-Kaon $\bar{K}$ = $\bar{s}d$

Kaon and anti-Kaon have parity -1 ($q\bar{q}$ has parity of +1 × -1 = -1).

C operator acts on Kaon, $C|K\rangle = |\bar{K}\rangle$ and $C|\bar{K}\rangle = |K\rangle$

$CP|K\rangle = -|\bar{K}\rangle$

This is not an eigenvalue equation as we want $-|K\rangle$ on the right side.

So, we need to combine Kaons and anti-Kaons to get form eigenvalue equations as only eigenvalues are measured in experiments.

$K_1 = \frac{1}{\sqrt{2}}\left(|K\rangle - |\bar{K}\rangle\right)$ has CP value of +1

$K_2 = \frac{1}{\sqrt{2}}\left(|K\rangle + |\bar{K}\rangle\right)$ has CP value of -1.

K_1 decays into two pions ($u\bar{d}$ quark combinations) and K_2 decays into 3 pions based on energy considerations. The decay times of K_1 and K_2 differ as well. K_1 has short decay time and is called K_S. K_2 has longer decay time and is called K_L.

$K_S \rightarrow \pi\pi$

$K_L \rightarrow \pi\pi\pi$

In 1964, Fitch and Cronin did an experiment where neutral Kaons are made to go through a tube. The Kaons were detected at short end and long end of the tube.

If CP was exact symmetry, we expect that K-short be detected at the short end of the tube and K-long to be detected at the long end as the K-short would have all decayed by that time.

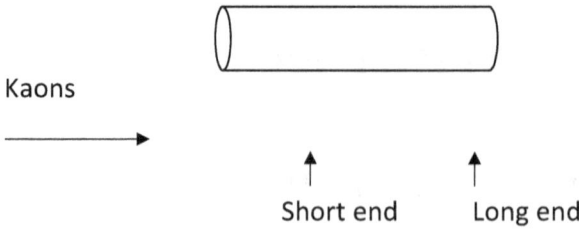

The Nobel prize winning experiment showed that $\pi\pi$ are detected at the long end as well. This is only possible if K_L (CP -1)→$\pi\pi$ (CP +1) happens, though rarely, violating CP symmetry.

There is an interesting question here. Is Kaon a particle or K_1 is a particle?

It depends on the situation. Kaons are produced in strong interactions but when CP is measured, K_1 comes into picture which is a linear combination of Kaon and anti-Kaon. It is like a person who is a basketball player. How do we know he plays basketball? We can see him playing with other players. He is also a voter, that can be detected by seeing him go to the polling booth. So, what a person or a particle is depends on what are we asking and measuring.

The Kaons have other decay modes as well.

$K \rightarrow \pi^+ e^- \bar{v}$

$\bar{K} \rightarrow \pi^- e^+ v$

So, K_L made of K and $\bar{K}$ should have equal probability of $\pi^+ e^- \bar{v}$ or $\pi^- e^+ v$.

The experiments show that decay to $\pi^- e^+ v$ is preferred. This means there is now experimental difference between matter and anti-matter. An electron can be defined as a particle found in lesser decay mode of long-lived Neutral Kaons.

CP violation alone is not enough to show why our universe is largely made of matter with so little anti-matter. The full answer is a still hotly debated.

CPT Theorem

Physicists never give up. The charge conjugation parity is violated, parity is violated, CP is violated too. Physicists have hope on CPT symmetry. The time symmetry is added to the CP symmetry. This symmetry means that the product of charge conjugation, parity and time reversibility should remain the same. This means a process where CP is broken, time reversibility should be broken as well to keep CPT unchanged.

By time reversibility, we mean if a process is

A + B → C + D then C + D → A + B is also equally likely.

The CPT theorem is based fundamental to quantum field theory as it implies locality and Lorentz invariance. The breaking of CPT will shake the foundations of QFT. So far, no experimental evidence exists that it is broken. Time will tell if this sacred symmetry survives the test of time.

The neutrinos only feel weak interactions. Beta decay is a classic example of weak interaction.

n → p + e + anti-neutrino

The weak interactions also change flavor, so up quark can become down quark.

The propagator formula is also slightly modified.

$\frac{g_w}{q^2 - M^2 c^2}$. Since the mass is quite large as compared to momentum q, the q term is ignored.

The scattering amplitude is

$\mathcal{M} = \frac{g_w}{M^2 c^2} \times \gamma^\mu (1 - \gamma^5) \times$ Energy difference × Momentum difference between incoming and outgoing particles.

Each process has unique formula by doing calculations of momenta and energy and summing over spins as described in various chapters. I am not going to list these formulae as it is largely an algebraic exercise.

Pion Decay

$\pi^- \to e^- + \bar{v}$. The pion (u and d quark) can decay into electron and anti-neutrino

OR

$\pi^- \to \mu + \bar{v}$. The pion can decay into muon and anti-neutrino (muon type).

It turns out the muon pathway is preferred even though electron is much lighter than muon and should have been the preferred route.

This curious case can be explained by the spin and handedness.

The pion is spin zero. The anti-neutrino is always right handed (being antiparticle).

To preserve angular momentum and to make total spin zero, the emitting electron will also be right handed. The weak interactions only involve left-handed particles. If electron was massless, it would be left handed, as massless object's handedness cannot be changed. Since electron has small mass, it is left leaning. The muon being heavier is right leaning. Since the emitting particle needs to be right handed, more right leaning muon takes the cake.

Flavor Change

There are certain rules that govern flavor change in weak interactions. The leptons change flavor only within a generation, there are is no cross mixing.

$\binom{v_e}{e}$ $\binom{v_\mu}{\mu}$ There is no mixing in these 2 generation of leptons.

The electron will only emit electron neutrino, not muon neutrino.

But in case of quarks, there is some mixing allowed.

$\binom{u}{d}$ $\binom{c}{s}$ $\binom{t}{b}$ There is mixing allowed, up quark can go to down quark but there is some probability of it changing to strange or bottom quark as well.

The new mixed states can be called d^{mix} and s^{mix}.

$d^{mix} = d \cos\theta_c + s \sin\theta_c$

$s^{mix} = -d \sin\theta_c + s \cos\theta_c$

They are linear combination of old states and cos, sin θ represent probability of finding each state.

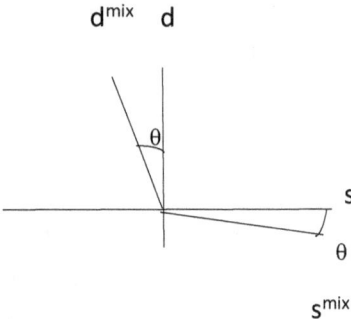

The new states can be thought of as rotated states from the original ones and the rotating angle θ is called Cabibbo angle.

$$\begin{pmatrix} d_{mix} \\ s_{mix} \\ b_{mix} \end{pmatrix} = (3 \times 3 \text{ matrix of } \theta s) \begin{pmatrix} d \\ s \\ b \end{pmatrix}$$

The 3 × 3 matrix is called CKM matrix. The values of matrix elements determine how much mixing is there. The values of the matrix come for experiments only. The standard model makes no theoretical prediction of these values.

The intuition behind this mixing and math is straightforward. Let me give you an example from the political world.

The political donors are of two types, Republican and Democrat. There is some mixing observed, with Republican donor giving money to Democrats and vice versa.

New R donor = Republican cos θ + Democrat sin θ.

Similarly, new D donor is also a linear combination of Republican and democrat donors.

The angle θ determines how much cross donation take place. This can be measures by political donation data.

If it was like standard model, you would say that there are no explanations why certain donors switch parties. The data or experiments tell us everything knowable about them.

This would be a view of a foreigner, who does not understand American political system and would be ignorant about the reason behind party switching.

An astute observer of American politics would be able to find reasons to explain the party switching. A coal mine union worker who is a traditional Democrat donor may have voted for a Republican who is anti-environment and wants to keep coal mines open. Similarly, a traditional Republican donor who is retiring may vote Democrat to keep his social security and pension payment intact.

This is an obvious weakness of the Standard Model. The mixing angles have no explanation except experimental verification. We need an astute observer to come to the Standard model and explain the mixing angles once for all.

Neutral Weak Interaction

The neutral interactions are mediated by Z boson unlike the charged ones which are W mediated. The neutral interactions are slightly different than charged ones.

The flavor change is not allowed so lepton and quarks do not change into one another.

The vertex factor is more complicated.

W boson factor is $g_w \gamma^\mu (1 - \gamma^5)$, ignoring certain constants and i's

Z boson factor is $g_z \gamma^\mu (c_v - c_a \gamma^5)$, each fermion has a distinct factor e.g. neutrino has c_v and c_a values of ½ and ½.

Electron has c_v and c_a values of -½ + 2 $\sin\theta_w$ and -½ and so on. There is a table you can refer to for each fermion.

The coupling factors are linked to one another by angle θ_w.

$g_w = g_e$(electro-magnetic)/ $\sin \theta_w$

$g_z = g_e / \sin\theta_w \cos\theta_w$

θ_w = 28.7 degree.

w is called Weinberg angle. Again, it is experimentally determined.

The Z boson couples to left and right-handed particles, but differently.

($c_v - c_a$) for left handed,($c_v + c_a$) is vertex factor for right -handed fermions.

The weak neutral interactions are hard to detect. They are overwhelmed by electromagnetic interactions.

Electron-positron scattering

$e^- + e^+ \rightarrow \mu^+ + \mu^-$

The electron and positron annihilate into virtual boson leading to production of muons. The process can take place via photons or Z bosons. The photons dominate at low energies. If we calculate the scattering cross section of each process and compare the strength, we will find

$$\frac{\sigma(z)}{\sigma(\gamma)} = 2 \left(\frac{E}{M_z c^2}\right)^4$$

Since the mass of Z boson is large, $M_z c^2$ term >>> 2E. The denominator is much larger, so the photon dominates. But at the Z pole, where 2E = $M_z c^2$, the ratio is close to 200. This means Z boson dominates at high energy. It is like saying the more agile car will be faster than a heavier truck but given enough power, the sluggish truck can give the sports car, a run for its money.

Electro- weak Unification

The unification of electromagnetic and weak interactions has been a remarkable achievement of the Standard Model. The credit goes to physicists like Steve Weinberg, Abdus Salam and Sheldon Glashow. They were awarded Nobel prize in 1979 for their efforts.

The first step in this process is to simplify the notation and absorb $(1 - \gamma^5)$ term.

ϕ_L (left handed) = $(1 - \gamma^5) \phi_P$

This we already know. A technical point here is that ϕ_L is a chiral state and it represents helicity of -1(by convention), only for massless and nearly massless particles. The massive particles can have different chirality and helicity as we can always change the direction of motion.

The factor $\gamma^5 \phi_P$ picks up the helicity factor

$\gamma^5 \phi_P$ = (momentum. spin) ϕ_P. If ϕ_P helicity is +1 then $(1 - \gamma^5) \phi_P = 0$

This is just mathematical way of saying that $(1 - \gamma^5)$ only works on a state that it is supposed to work!

Likewise $(1 + \gamma^5)$ is the right- handed factor. The antiparticles have opposite sign to their corresponding particles e.g. anti-particle of left-handed particle has $(1 + \gamma^5)$ factor.

Next step is club left handed particles into groups and use the mathematics of spin to unite them.

$\zeta_L = \begin{pmatrix} v_e \\ e \end{pmatrix}_L$

The particles can be described by spin states $\begin{pmatrix}1\\0\end{pmatrix} \begin{pmatrix}0\\1\end{pmatrix}$. This is not real spin but just a way to describe the particles.

The spin matrices do the job of picking the right particle.

$$\tau^+ = \begin{pmatrix}0 & 1\\0 & 0\end{pmatrix} \quad \tau^- = \begin{pmatrix}0 & 0\\1 & 0\end{pmatrix}$$

Then we form a current which is just these lefties marching together.

$J^+ = \bar{\zeta}_L \gamma_\mu \tau^+ \zeta_L$, this is weak charged current.

It means that $(\bar{\nu}_e \; \bar{e}) \begin{pmatrix}0 & 1\\0 & 0\end{pmatrix} \begin{pmatrix}\nu_e\\e\end{pmatrix} = \bar{\nu}_L \gamma_\mu e_L$

Similarly, the second current is J^-.

This is like total spin +1 and -1. It will take care of W^+ and W^- bosons.

Obviously, there is the third combination possible, spin zero (singlet state).

This will take care of the neutral Z boson. It is called isospin-3, like the z component of the spin.

$$\tau^3 = \begin{pmatrix}1 & 0\\0 & -1\end{pmatrix}$$

The current formed is called neutral current, j^3.

$J^3 = \bar{\nu}_L \gamma_\mu \nu_L - \bar{e}_L \gamma_\mu e_L$, remember the bar on top represents outgoing particle.

It is like neutrino-electron scattering.

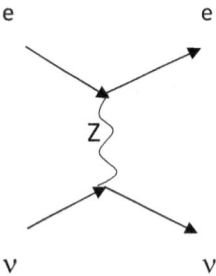

We have incorporated W and Z bosons into our so-called weak spin framework but there is a problem. The Z boson acts on right-handed particles too but the J^3 current is only left handed. The photon is also missing.

Not surprisingly, we form the fourth current that includes photons and Z bosons.

But first, the electromagnetic current is $J^{em} = - \bar{e}_R \gamma_\mu e_R - \bar{e}_L \gamma_\mu e_L$. This current has both R and L particles as photons do not distinguish between handedness.

Then, we combine J^{em} and J^3 to form the combined current called weak hypercharge current.

$J^Y = 2J^{em} - 2 J^3$. This current has combined photons and Z boson. It preserves the property that Z boson cares about left and right handedness, but photon does not.

What is Y?

It is a number assigned to each fermion.

Where there is charged current, there should be a charge. Now, it is not ordinary electric charge as Z boson is mixed in. It is called weak hyper charge.

Y is used to calculate weak hypercharge.

Q (hyper charge) = I^3(weak iso spin) + ½ Y

Let's calculate the hyper charge for electron.

Iso- spin for left handed electron is - ½. Minus sign is convention. This comes from forming pair of electron and neutrino as explained above. $\begin{pmatrix} \nu_e \\ e \end{pmatrix}$ to $\begin{pmatrix} 1 \\ 0 \end{pmatrix} \begin{pmatrix} 0 \\ 1 \end{pmatrix}$.

½ is to normalize.

The right-handed particle has iso-spin zero as right handedness is excluded from iso -spin. Y number for e_L is -1 and e_R is -2.

Q of e_L = -½ + ½ × (-1) = -1

Q of e_R = 0 + ½ × (-2) = -1.

Thank God! The left and right-handed electron has same charge of -1. This is what we expected as outside of weak interaction, there is no need to distinguish between left from right-handed electron. The electric or weak hypercharge of electron is -1.

GWS Model

Glashow-Weinberg- Salam model combines the electro-weak forces albeit with more mathematics.

j.W + j^Y B

W has three current streams W^1, W^2 and W^3. The familiar charged W bosons are linear combinations of those.

$W^+ = W^1 - iW^2$, $W^- = W^1 + iW^2$

j.W = $J^1 W^1 + J^2 W^2 + J^3 W^3$

The first two are the charged W bosons and the third term W^3 forms Z boson.

The W^3 along with B bosons form linear combinations to form Z boson and photon.

Z = $W^3 \cos \theta_w$ − B $\sin \theta_w$ and γ(photon) = $W^3 \sin\theta_w$ + B $\cos \theta_w$

Where θ_w is the Weinberg angle

The angle is purely experimental and there is no explanation in the Standard Model for its value. Like Cabibbo angle, it is used to get photon and Z boson from the above linear combinations.

$$\begin{pmatrix} \gamma \\ Z \end{pmatrix} = \begin{pmatrix} \cos & \sin \\ -\sin & \cos \end{pmatrix} \begin{pmatrix} B \\ W \end{pmatrix}$$

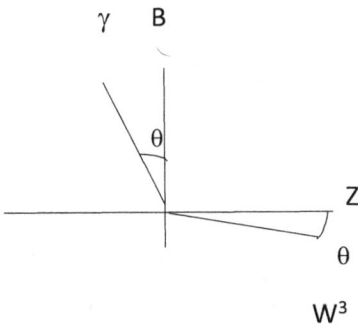

Do you remember looking through polaroid glasses? When you look at certain angle, certain colors look clearer, at another angle they become blurred again. Try it while looking at your car's heads up display.

The Weinberg angle can be thought of like that. At certain angle, you see photon and Z boson, at another angle, they become B and W boson. You do not need polaroid glasses but Professor Weinberg's intuitive vison to see it!

The question how Z boson gets a mass and photon is massless is even more complicated. This needs introduction to spontaneous symmetry breaking and Higgs mechanism. We will discuss these in detail in subsequent chapters. Hold on to your seat belts.

Chapter 14

Quantum Field Theory

The physicist's first attack weapon is to find a Lagrangian. This is like hunting for the right talent for your job. Lagrangian as you may recall is KE − PE. This is how nature uses its reserves judiciously. In QFT, Lagrangian is written as your best guess. The Lagrangian is then used to derive the equation of motion or Euler -Lagrange equation, which tells us how fields or particles move and interact.

In QFT, particles are represented by fields. Let's start with scalar field ϕ

The Lagrangian for bosons (spin 0) is

$$\mathcal{L} = \frac{1}{2}\partial_\mu \phi \partial^\mu \phi - \frac{1}{2}\left(\frac{mc}{\hbar}\right)^2 \phi^2$$

$\frac{\partial}{\partial t}\left(\frac{\partial L}{\partial v}\right) - \frac{\partial L}{\partial x} = 0$ is Euler Lagrange equation to get equations of motion.

Here we replace $\partial t(\partial v)$ with $\partial(\partial \phi)$ and ∂x with $\partial \phi$.

In classical Lagrangian

$\partial v \left(\frac{1}{2}v^2\right) = v$ and $\partial t(\partial v) = a$

Velocity or $v = \frac{\partial}{\partial t}$ and acceleration $a = \frac{\partial}{\partial t^2}$

So, in relativistic notation, after doing differentiation with $\partial(\partial \phi)$, we get

$\partial_\mu \phi \partial^\mu \phi$ which is nothing but $\frac{\partial}{\partial t^2} - \frac{\partial}{\partial x^2} - \frac{\partial}{\partial y^2} - \frac{\partial}{\partial z^2}$

After doing differentiation, we get the equation

$\partial_\mu \phi \partial^\mu \phi + (\frac{mc}{\hbar})^2 \phi = 0.$ This is the Klein-Gordon Equation.

Mission accomplished, this is what we were hoping to get.

Similarly, Lagrangian for Dirac equation (fermions)

$\mathcal{L} = i\hbar c \bar{\psi} \gamma \partial \psi - mc^2 \bar{\psi}\psi$

Applying Euler-Lagrange equation and differentiating with respect to $\frac{\partial}{\partial \psi}$, we get back the Dirac equation

$i\gamma \partial \psi - \left(\frac{mc}{\hbar}\right)\psi = 0$

Here ψ is a spinor field, it carries spin and represents fermions unlike the spin less bosonic ϕ field.

Let's learn about gauge invariance in QFT as it's at the heart of the theory.

It is a very technical and mathematical subject devoid of intuition. I will nevertheless try to get the basic concepts across. There are two types of invariance that we need to know- global and local.

Global gauge invariance- Here a phase factor is added to the field and it cancels out with the opposite sign of the complex conjugate phase. In simple words, we do something to the whole system at once, it implies that it should not affect the system. It is typically some kind of rotation. Remember fields are complex objects and rotation does not occur in physical coordinates but in the imaginary complex space.

$e^{i\theta} \phi$, phase is constant as θ is constant

$e^{-i\theta} \phi^*$ is the complex conjugate of the above phase.

The probability does not change as $\phi^*\phi$ remains the same, $e^{i\theta}$ and $e^{-i\theta}$ cancel.

The differentiation is not affected as $e^{i\theta}$ is constant term so $\partial(e^{i\theta}\phi) = e^{i\theta}\partial\phi$.

In layman terms, let's say you are traveling on the highway and going home. The earth is rotating. So, everything- your car, house and road is being rotated simultaneously everywhere. This is global invariance as earth's rotation has no influence on your trip to home.

Local invariance- What if θ is not constant, so the phase factor changes from place to place. $e^{i\theta(x)}$ is variable and depends on coordinate x.

$$\partial(e^{i\theta(x)}\phi) \neq e^{i\theta(x)}\partial\phi$$

By rules of product differentiation, answer now will be

$$\partial(e^{i\theta(x)}\phi) = e^{i\theta(x)}\partial\phi + \phi\,\partial(e^{i\theta(x)}) \text{ extra term}$$

This extra term is a nuisance and needs to be get rid of.

So, subtract it from the Lagrangian to make it invariant.

Locally invariant $\mathcal{L} = \mathcal{L}$ (with extra term) – extra term

This is done by defining a new type of derivative called covariant derivative.

It removes the extra term while doing the differentiation.

D (covariant) = ∂ + extra term

It is defined this way because local phase factor is usually $e^{-i\theta(x)}$. So extra term comes with a negative sign and we add positive term to cancel it.

What does this mean intuitively?

Let's again take the case of you traveling in the car to your home. As you move uphill or downhill (local change in the steepness of road), the car speed changes. In other words, local changes in the road DO affect the car speed. In order to maintain car speed (local invariance), you have to counteract with your throttle and break.

This way you maintain the car speed and produce local speed invariance. The throttle and break application are like covariant derivatives that cancels out the effects of local gauge changes.

In quantum mechanics things are not as simple as it seems. Since we are looking for theory of scattering of particles, we have to account for propagators like photons in the Lagrangian. We introduce a gauge field which is a vector field.

It interacts with the field ϕ, so there is an interacting term. It also has to exist as a free field, so there is a free field term.

It is like saying that while traveling by car, there is air interacting with the car, then there is free air as well, not directly interacting with the car.

We call this field A. The interacting term will be like $\bar\phi\phi A$. The free field is written as $F^{\mu\nu} F_{\mu\nu}$. F is made of derivatives of A, $\partial^\mu A^\nu - \partial^\nu A^\mu$. This is all electromagnetic theory. A is electromagnetic potential. The derivates of electromagnetic potential give electric and magnetic fields. We can of course make changes to A without affecting the magnetic field.

Recall magnetic field B = curl of A, $\nabla \times A$

If we add a gradient to A, grad λ, then $\nabla \times \nabla\lambda$ is zero. Curl of a gradient is always zero. Gradient means going uphill or downhill, but curl needs rotating motion. The gradient cannot reach the same point it started from in a circular fashion as required by the curl, so it is incompatible with a curl.

This is the gauge invariance, so adding gradient $\partial\lambda$ has no effect on the field.

The phase factor θ can be replaced by $\partial\lambda$. So extra term will have A, instead of θ.

The global and local gauge invariant Lagrangian is

Original Dirac Lagrangian – free field – extra interacting field term

$i\hbar c\bar\psi\gamma\partial\psi - mc^2\bar\psi\psi$ - $F^{\mu\nu} F_{\mu\nu}$ - $q \bar\psi\psi\gamma A$

I have left some factors of π and μ(space time- t, x, y and z) scripts to keep things relatively simple.

The phase factor multiplication $e^{i\theta}\psi$ is like multiplying $U\psi$. U is a unitary matrix.

The mathematical jargon is U's are part of U (1) group. UU* is 1.

Basically, multiplying by 1, does nothing. This is what U (1) group means. It does nothing to the lagrangian as $1 \times \mathcal{L} = \mathcal{L}$.

A lagrangian which is globally invariant is good, the one which is locally invariant is even better. The theory that does not change when we do things to it is a good thing. Back to our car analogy, we do not want earth's rotation(global) or local factors like air interacting (gauge fields) with the car to affect our journey. We employ strategies like aerodynamics of the car, breaking, throttle (covariant derivatives) to cancel these effects and finish our journey on a happy note.

Yang Mills Theory

Yang -Mills theory takes the concept of global and local invariance and gives a powerful framework on which the skyscraper of Standard Model is built. We have seen phase or U (1) symmetry being incorporated into the invariant Lagrangian.

SU (2) and SU (3) have to be included as well.

To achieve this, we need to add more things to the equations. First of all, ψ can have sub components like $\psi_1\psi_2$. After all, there are more than one particle or field and they interact and scatter. In our car analogy, you are not alone on the road, you have the damn traffic.

The phase θ can be more complicated as well. The general expression is e^{iH}. Here H is Hermitian which means observable or measurable.

$H = \theta + \tau$. τ is our familiar 2×2 matrix representing spin. This way SU(2) can be included in the theory.

Again, the invariant Lagrangian has the original Dirac Lagrangian but it has 2 fermion fields ψ_1, ψ_2. The free field is made of derivatives of gauge field A. The extra term is modified and has interaction term with Dirac and gauge fields.

Extra term = $(q\,\bar{\psi}\psi\gamma)\cdot A$. Here q is coupling constant like electric charge. The field A is of 3 types and we have to take dot product with each of 3 Pauli matrices.

The Pauli matrices do not commute with each other (uncertainty principle, cannot find x, y and z spins together). So, $\tau \cdot A$ is made of 3 components that do not commute. Technical name for this is non-Abelian gauge theory as opposed to U(1) group which is commuting and is called Abelian gauge group.

The gauge field is massless as mass messes up local and global invariance.

Finally, let's make this theory colorful. The three colors-red, blue and green of QCD are added to the Dirac field. ψ is now made of $\psi_{red}, \psi_{blue}, \psi_{green}$. The massless gauge field is made of 8 components. The Pauli matrices are replaced by Gell-Mann matrices (λ's).

$\mathcal{L}$ = (3 Dirac color fields) – Free Field – extra interacting term $\lambda \cdot A$.

A is the source of the free field F. The quanta of the field F are 8 gluons.

Summing up, we are not alone on the road, there are other cars (Dirac fields).

The air is not clean as well, it has dust, water, oil etc. (8 gluons) that interact with

cars and exist freely as well. The cars navigate through the traffic and the effects of air, dust is minimized by appropriate measures (gauge invariance).

Higgs Mechanism

We have a big problem. The massless gauge fields can explain interactions by photons and gluons as both are massless but fail to explain weak interactions through W and Z bosons as they carry mass.

This needs introduction of a new concept called Spontaneous Symmetry Breaking, SSB.

The concept is easy to understand. If there are bunch of apples sitting on the table and they are all similar, we have a symmetry. After few days, we will find that a random apple will become rotten. This happens spontaneously, and it break the symmetry of apples. This is spontaneous symmetry breaking. If we put one apple in the refrigerator and another one outside, obviously the outside one will become rotten first. This is explicit symmetry breaking. The atoms in a metal may spontaneously prefer a direction based on a random selection of spin by one individual atom.

Let's introduce some mathematics. Start with a boson field ϕ. It can be made of real and imaginary parts. $\phi = \phi(real) + i\phi (imaginary)$.

The lagrangian would be $\mathcal{L}$ = KE -PE

KE= ½ $\partial \phi \partial \phi^*$

For the potential, we chose a special function, that breaks spontaneous symmetry. This means it prefers a certain configuration and is unstable at the origin. The physicists call it a Mexican hat potential.

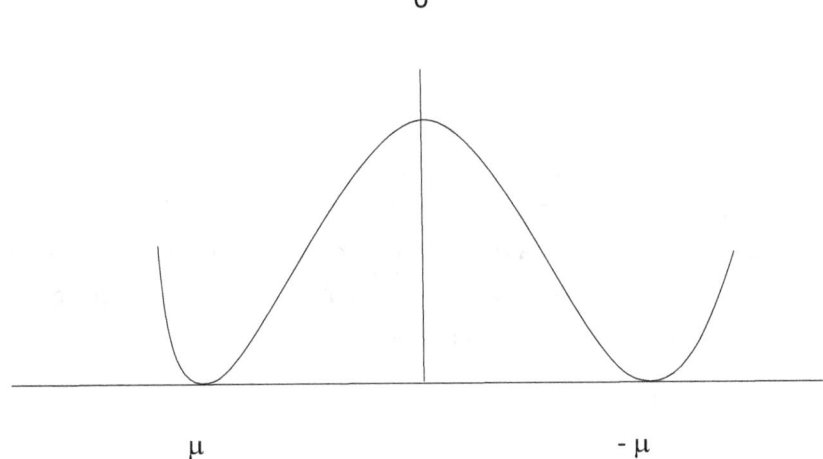

The potential is not minimum at zero, but it prefers either μ or $-\mu$. It is like trying to stabilize ball on a hat, it will roll down from tip of the hat to the brim and may rotate around the brim.

A potential term would look like $\mu_2 \phi^2 + \lambda^2 \phi^4$ + higher order terms. This is like saying potential value is 2.15 and first term approximation is 2, 2nd term is 0.14 and so on. $\lambda^2 \phi^4$ is also called self-interacting term. Our weight is made of body mass and clothes. The self-interacting term would be like clothes which you can shred and wear.

The Lagrangian now looks like

$\mathcal{L} = \frac{1}{2} \partial\phi\partial\phi^* - \mu_2 \phi^2 + \partial\xi\partial\xi^*$ + interaction term ξ, ϕ

ξ is like A, gauge field but again, its massless. This is a disaster, we want the gauge bosons like W and Z to have mass. This was the whole point of introducing spontaneous symmetry breaking.

Mass Recognition

It is important to know how to recognize mass and massless terms.

The terms with derivatives are always massless e.g. $\partial \xi \partial \xi^*$ or $\partial \phi \partial \phi^*$. The mass term is typically like $\mu_2 \phi^2$ or $m^2 \Psi$. Think of ϕ as a car. If you want to weigh the car, you take the whole car and put it on the scale. The value you get is weight of the car or $m^2\Psi$ like term. $\partial \phi \partial \phi^*$ is like a moving car, ∂ represents change. The moving car cannot be put on a scale! So, it represents a massless term.

Where does $\partial \xi \partial \xi^*$ come from and what does it represent?

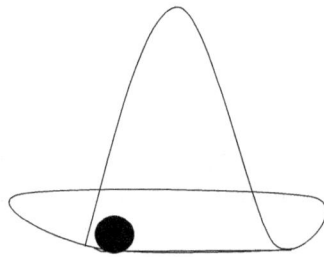

$\partial \xi \partial \xi^*$ represents the movement along the brim of the hat and the quanta of this field is the massless Goldstone boson.

The final step in giving gauge bosons a mass is to combine local invariance and spontaneous symmetry breaking.

Recall that when we added a phase $e^{i\theta(x)}$ to ϕ, we had to introduce covariant derivative to soak up the extra term.

D (covariant) = ∂ + extra term(iA)

D* (complex conjugate) = ∂ - extra term(iA)

Putting these derivatives in the Lagrangian, terms like ½ $\partial\phi\partial\phi^*$ will be replaced by $(\partial + iA)(\partial - iA)$. If you carry out the multiplication of all the terms in the expression, there will be term like $\mu_2 A^2$. This is the mass term for the gauge field A. Now we are in business! The gauge field A has acquired a mass by using local gauge invariance and derivatives. The Lagrangian now looks like

$\mathcal{L}$ = ½ $\partial\phi\partial\phi^*$ - $\mu_2 \phi^2$ + $\partial\xi\partial\xi^*$ + FF (derivatives of A) + $\mu_2 A^2$ + interaction terms of A,ξ,ϕ.

$\partial\xi\partial\xi^*$ term can be eliminated. Recall the field ϕ is made of real and imaginary parts $\phi_1 + i\phi_2$. If we chose a particular value on the brim of potential, then ϕ_1 has a particular say μ, then the other part ϕ_2 represents ξ as ball can move from μ to around the brim represented by ξ.

To eliminate ϕ_2

Chose a new field $\phi_{new} = e^{i\theta}\phi$.

$e^{i\theta}$ = cos θ + i sin θ

So, ϕ_{new} = (cos θ + i sin θ) (ϕ_1 + i ϕ_2) or (ϕ_1 cos θ - ϕ_2 sin θ) + i (terms)

θ can be chosen cleverly to make ϕ_2 zero and ϕ_{new} real only.

This way $\partial\xi\partial\xi^*$ term is eliminated.

I know this is a lot of mathematical jugglery but basically if you pin down the ball rolling freely on the brim, the ball cannot move. The moving degree is eliminated. If ball cannot move, then the Gold stone boson which is the result of moving of ball around the rim is eliminated as well. This happened because the potential took a particular value μ by symmetry breaking and the use of covariant derivatives did the rest. The physicists describe this process rather poetically that the Goldstone boson got eaten by the gauge field and it gained mass as a result.

In summary, the whole universe is filled with Higgs field. The particles that interact with it gain mass (W, Z bosons, fermions etc.). The particles that pass through it unaffected are massless (photons, gluons). The field itself has quanta called *Higgs boson* or scandalously called God particle.

The God has created Higgs road everywhere. Luckily it is pull of pot holes and craters. The cars that interact with the road fall into craters, feel the resistance and thus gain mass. The things that fly, do not interact with the craters on the Higgs road and remain massless. The craters themselves get chipped away with cars hitting them all the time, creating splinter particles (Higgs boson).

The discovery of Higgs boson at the Large Hadron Collider in 2012 was the last missing piece in the Standard Model. It was a monumental achievement and an acknowledgement of the power and prediction of the Standard Model and underlying Quantum Field Theory.

It is worth noting that not all particles get mass from Higgs phenomenon. In fact, most of the ordinary matter, protons and neutron get their masses from interactions of gluons inside them like a hot glass of soup with high energy particles bumping into one another can create mass.

In case you are wondering how to create Higgs boson in the lab, the answer is through proton-proton collisions. There are many processes that take place in these collisions. The processes involving gluon fusion and creation of heavy quarks lead to Higgs boson production. The heavier the particle, the strongly it interacts with the Higgs field, leading to Higgs boson production. In our Higgs road analogy, the bigger vehicle will interact strongly with the crater on the road. The jolt from the bigger vehicle causes the crater to spit out dust or Higgs boson.

Planck Units

The SI units of kg, meter and second are not well suited for quantum mechanics. These units are for human consumption and they are huge in comparison to scales of a quark or neutrino. Then there is the problem of constants in equations. There are $\hbar$, c 's all over the place in equations. The physicists have come with a clever idea to set these constants of nature =1. This simplifies lots of equations and gives us Planck units which are more natural to the quantum world.

Planck constant $\hbar = 1$.

Speed of light $c = 1$.

Gravitational constant $G = 1$.

$E = \hbar f$ will become $E = f$

$E = mc^2$ will become $E = m$ and so on.

This simple step is quite illuminating. You can see energy is equivalent to mass and frequency in these units.

Let's derive Planck units. You use dimensional analysis which is a very clever technique to check if equation makes sense.

Let's take $E = mc^2$

Units of energy is Joule, which is $kg \left(\frac{m}{s}\right)^2$. The speed of light c has $\left(\frac{m}{s}\right)^2$ units so mc^2 units match E.

$[M]^1 [L]^2 [T]^{-2} = [M]^1 [L]^2 [T]^{-2}$ in dimensional analysis jargon.

To get Planck length, we need to know how constants are dimensionally related

$L^P = G^a \hbar^b c^c$

$L^P = [M]^0 [L]^1 [T]^0$

We have to put back units of constants in the right-side equation and make 3 equations of M, L and T and equate both sides to get the value of a, b and c.

Without actually solving the equation, let me just give you the answer.

$L^P = \sqrt{\dfrac{\hbar G}{c^3}}$ which is equivalent to 10^{-35} m in SI units.

Similarly, Planck mass and time can be calculated.

Is there more to Planck units than just convenience?

Many physicists believe that Planck length is the smallest length scale and it could signify quantization of space. The Planck time is the time light takes to travel Planck length. Time (bigger than Planck scale!) will tell whether this is true.

Neutrino Problem

Neutrino is the problem child of the Standard Model. Neutrinos come in three flavors- electron, tau and muon. The Standard Model predicts that neutrinos are massless. They are electrically neutral and take part in weak interactions. The neutrinos exist only as left-handed particles and their anti-particles or anti neutrinos are right- handed. Being massless, there is no mixing allowed. The massless particles always move with the speed of light. In their frame, time is still and does not change. So, no change is allowed as the time is frozen. The massless particles cannot be slowed and thus their handedness is fixed as well.

The problem comes from experiments that predict neutrinos have a very small mass and thus cannot move at the speed of light. The solar neutrino problem clinched the argument in favor of mixing of neutrino flavors. Sun is an important source of neutrinos due to nuclear reactions. The amount of electron neutrinos detected were less than predicted by calculations based on nuclear fusion. It means that electron neutrinos convert into tau and muon neutrinos, which were not detected by experiment.

In quantum mechanics how do you describe this process?

We make superposition of flavors and use matrices to mix them up.

$N_1 = \cos\theta \, \nu_\mu - \sin\theta \, \nu_e$

$N_2 = \sin\theta \, \nu_\mu + \cos\theta \, \nu_e$

The mixing angles are parameters determined by experiments.

In sports analogy, a typical Canadian sports fan is a mixture of hockey and say baseball.

Sports fan = $\cos\theta$ hockey + $\sin\theta$ baseball

Mixing angle θ can depend on the weather, favorite team and friends. So, the probability of being a baseball fan is high in the summer or if your favorite team is winning.

The experiments show that Torontonians tilt their mixing angle towards baseball whenever Toronto Blue Jays enter playoffs!

The probability of getting a particular flavor from the superposition can be put into a matrix form.

$$\begin{pmatrix} \nu_e \\ \nu_\mu \\ \nu_\tau \end{pmatrix} = \begin{pmatrix} U & U & U \\ U & U & U \\ U & U & U \end{pmatrix} = \begin{pmatrix} N_1 \\ N_2 \\ N_3 \end{pmatrix}$$

This is like putting three sports into the mixture and there will be lot more combinations to keep track.

Chapter 15

Beyond Quantum Mechanics

Unification

Unification has long been a cherished goal of physicists. The premise that all fundamental laws of nature arise from one basic principle or theory is very appealing. It is like monotheistic religion where idea of one God is very powerful. The unification of physics is nothing new. The modern physics began with Newton mechanics. He unified the motion of celestial bodies with the fall of apple on the ground. The Newton's laws of motion apply to everything in the universe. James Maxwell unified electricity and magnetism through Maxwell's famous equations. In early 20th century, two revolutionary theories developed independently. General and Special Theory of Relativity deals with cosmological scale, leading to the unification of space and time into one space-time continuum. Quantum Mechanics is the dominant theory of elementary particles. The unification of Quantum Mechanics and Special Relativity led to the development of Quantum Field Theory. QFT further refined the understanding of elementary particles and led to the formation of the Standard Model. Quantum Electrodynamics (QED) was developed first and was very successful in explaining electron behavior and photon interactions. It took longer to develop the theory of strong and weak interactions. The successful explanation of forces in the nucleus led to the formation of Quantum Chromodynamics (QCD) and weak interactions. QED and weak interactions were unified further into electro-weak theory. The unification of QCD and electro-weak theory still remains elusive. The other elephant in the room is gravity, which remains the most difficult to reconcile with QFT. General Relativity or gravity is a hard nut to crack and its incorporation into QFT leads to non-renormalization theories, which means infinities or in layman's terms nonsense!

Grand Unified Theory (GUT)

No GUTs No Unification! The GUTs try to unite strong force with electro-weak forces.

In unitary group jargon, SU (3) with 3 colors is unified with SU (2) and SU (1) representing isospin and photons. Many versions of GUTs exist but SU (5) or Georgi-Glashow model is the simplest and cleanest. 5 × 5 matrices can be used to mix up particles. Baryon number is not conserved which means quarks can be converted into electrons. Particles can be further sub grouped, which are called irreducible representations. The five particles usually group together, and it is called 5 representation. It includes neutrino, electron and down bar quarks with 3 colors.

$(v, e^-, \bar{d}, \bar{d}, \bar{d},)$ form our fantastic 5 group. U matrix acts on the group.

$$\begin{pmatrix} U & U & U & U & U \\ U & U & U & U & U \\ U & U & U & U & U \\ U & U & U & U & U \\ U & U & U & U & U \end{pmatrix} \begin{pmatrix} v \\ e^- \\ \bar{d} \\ \bar{d} \\ \bar{d} \end{pmatrix} = \text{mixing of particles}$$

There is a $\bar{5}$ representation where complex conjugate of the above particles is included- (v, e^+, d, d, d). Why form such a combination? It's mostly mathematical trickery that produces the desired theory. If we combine the electric charges of the group, they add up to zero. In other words, unitary matrices that act on them have trace zero.

$$\begin{pmatrix} v \\ e+ \\ d \\ d \\ d \end{pmatrix} \otimes \begin{pmatrix} v \\ e+ \\ d \\ d \\ d \end{pmatrix} \text{ or } \bar{5} \times \bar{5} \text{ leads to the remaining particles of the standard}$$

model. It is called $\overline{10}$ representation.

The gauge bosons convert one particle to another. We already know about gluons that convert one quark to another e.g. W, Z bosons and photons convert leptons into one another. Additional gauge bosons are needed for new interactions in this model. This is done through hypothesizing X and Y gauge bosons, which mix quarks and leptons. These gauge bosons live in a matrix form called the ad joint representation. X, Y bosons have color and electric charge and there are 6 sub types of each.

These particles have the same mass and interactions, which is not reality as strong and weak forces are very distinct. Higgs mechanism is used to spontaneously break the symmetry via vacuum having a non-zero value. This way we get back the Standard Model.

The most striking prediction of this model is the decay of proton. If it decays into leptons, it should be experimentally observed. The half-life of this process is very long, to the order of 10^{33} years. This is expected as most of the us are made of protons and we still exist! This long half-life can still be tested by having large collections of protons (Hydrogen) in big containers and detect the proton decay. Unfortunately, decades have gone by and there is no evidence for either proton decay or existence of X and Y bosons. The proton decay should not be confused with radioactive decay in which proton converts into neutron with the emission of positron and neutrino. This happens in unstable nuclei with excess protons in which quark exchange happens. It is called beta + decay. The baryon number is conserved in the process. A proton decay refers to a bare proton decaying, which has never been observed.

Even though experimentally GUT has been a dud so far, physicists still have some faith. A sceptic may ask what faith has to do with physics? A theory may be elegant and mathematically beautiful but if nature did not select it then it only belongs to the garbage bin. One reason for the faith is the coupling constants that govern interactions of SU (3), SU (2) and U (1) are not constants, but their value change based on the energy scale.

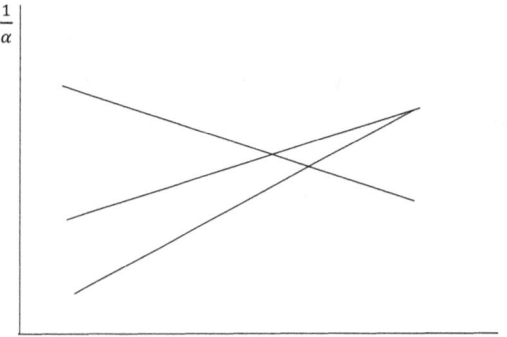

Energy Scale

Coupling constants (α) of SU (3), SU (2) and U (1) in the Standard Model cross each other at high energy but not quite at the same point. The magic of unified theories especially supersymmetry that we will study later, will be to make them converge at a single point.

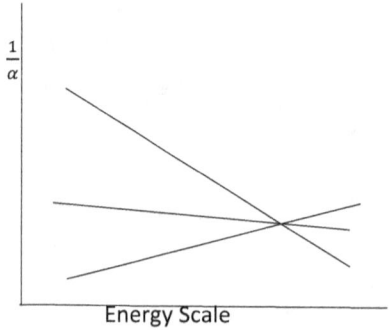

Energy Scale

This convergence takes place at around $10^{16}\ GEV$. Is this a coincidence or is it saying something really fundamental about nature? As you go to higher and higher energies, things look the same. It is like saying that we as humans have different beliefs, political ideologies etc. but if we go back in time, these distinctions disappear. Can you find any ideology in an infant?

Let's take a layman's GUT feeling on this theory. Sport fans come in various flavors. Americans from SU (3) with basketball, baseball and football.

Brits are SU (2), soccer and rugby. Finally, Canadians are U (1)- ice hockey fans. We can imagine these fans interacting with each other and even become fan of a new sport. A rugby fan may find NFL appealing as he will recognize the game easily. The sports gauge boson facilitating it could be a big glass of American beer!

Supersymmetry

Need is the mother of invention. The need to renormalize the mass of particles like Higgs boson led to the idea of supersymmetry. This idea is not a romantic deep thought experiment like apple falling from a tree or gedanken (thought) experiment of Einstein, where a man jumps from a building. It is a very technical and mathematical concept and is quite abstract. The concept of supersymmetry has permeated many fields of theoretical physics. Even string theory uses this concept.

Let's first look at the need to invent supersymmetry in the first place.

There are mass terms in the Lagrangian of the Higgs boson. They go like...

$$m^2\phi^2 + \mu\phi^3 + \lambda\phi^4\ldots\ldots$$

In terms of Feynman diagrams, these terms represent

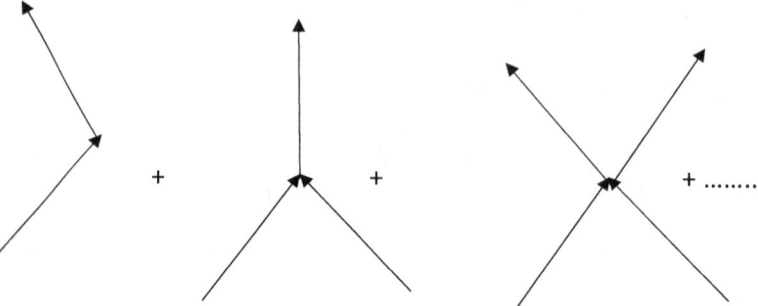

Let's look at an individual diagram, say ϕ^3 which represents 2 particles coming in and colliding to form one particle at a single vertex. The probability of that happening depends on the coupling factor μ. Even a single diagram can be made of lot of combinations as all possibilities have to be considered. The loops can be drawn on the diagrams where particles are created and annihilated at very close distances. These so-called self-energies lead to undesired infinities.

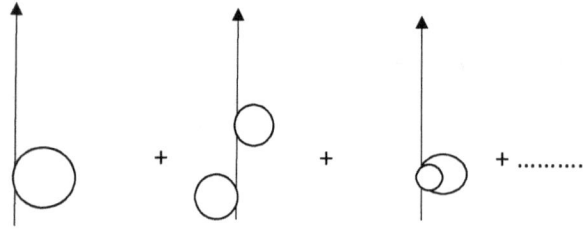

The contributions from these loops go like

$$\frac{\mu}{\Delta^2} + \frac{\mu^2}{\Delta^2} + \frac{\mu^3}{\Delta^2}\ldots\ldots$$

The Δ represents distance at which creation and annihilation takes place. It cannot take place at the same point otherwise as Δ goes to zero, the term goes to infinity. The Δ is a very small distance and but it contributes bigly to the term. The mass of Higgs boson is known from experiments and these divergent terms need to reconcile with experiment results. In some books Δ is replaced by momentum k. It is a common theme in physics to interchange distance and momentum through Fourier analysis as they represent the same information but in different domains like x ray or ultrasound of a body part shows the same information but with different emphasis.

The heart of the problem is this, say something weighs 10kg, it is made of various parts and their contributions are

9.99+ 10+ 100+ 1000+ 10000…. = 10kg

This is a problem and it requires fine tuning or fudge factors to reduce these contributions. It is not a satisfying solution as no fundamental principles are involved.

This is where supersymmetry comes in. The underlying principle is that every particle has a super partner. It means a boson has a fermion super partner.

What would that do?

The main use of this concept would be if somehow the super partner cancels the divergent terms. This is exactly what happens in supersymmetry.

The boson loop has a positive sign and fermion loop has a negative sign. Why is that? It is the same story of Pauli exclusion principle, meaning when fermions are exchanged, you get a negative sign. Similarly, if two fermion loops are exchanged, negative sign should be placed.

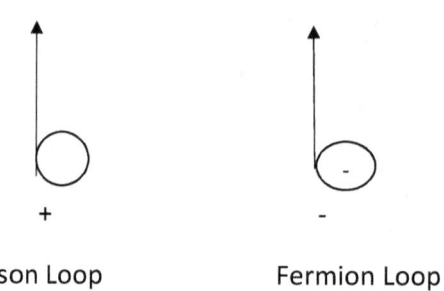

Boson Loop   Fermion Loop

Fermion loops contributions go like $-\frac{f}{\Delta^2} - \frac{f^2}{\Delta^2} - \frac{f^3}{\Delta^2}$. These will cancel the positive contributions from boson loops and we can have a sensible theoretical mass that matches experiments. Note that coupling factors µ and f should be similar for exact cancellations. This is the basis of supersymmetry.

SuperAlgebra

Supersymmetry theory has a bit of an ego problem, everything comes with super prefixes. The algebra used is called super algebra. It is the part of a subject called exterior algebra developed by German mathematician Hermann Grassmann. It was developed in the 19th century, long before anyone ever thought about supersymmetry. The Grassmann numbers are used extensively in supersymmetry.

Grassmann number is usually denoted by θ. It has a special property that if you interchange one Grassmann number with another, it changes sign.

$\theta_a \theta_b = -\theta_b \theta_a$

Or $\theta_a \theta_b + \theta_b \theta_a = 0$ or in other words they anti commute $\{\theta_a \theta_b\} = 0$

This is a useful property as we want something like this as fermion loop also gets a negative sign with interchange. These θ s don't represent fermions per say, they are just used to show fermionic properties. There are lot of properties that are associated with this algebra, I will list only few that are important to understand supersymmetry.

θ is usually a complex number, it can have various spin indices like left or right spin. θ can be considered like an odd number. So, even number times θ is an odd number and θ times θ is even.

$\theta^2 = 0$ is a useful property. If you are forming a polynomial equation like $a + b\theta + c\theta^2$, it ends at $b\theta$ as θ^2 term is zeros. It simplifies the equations. Now you can have term like $c\theta_1 \theta_2$ as θs are different. In this case $\theta_1 \theta_2$ term is not zero.

There are some peculiar properties of differentiation and integration.

$\frac{\partial}{\partial \theta_2}(a + b\theta_1 + c\theta_2 + d\theta_1 \theta_2) = 0 + 0 + c - d\theta_1$

Most of the steps are straight forward differentiation but the last step involves going through θ_1 which results in a minus sign. If it was $d\theta_2 \theta_1$ then the sign would remain positive. Keeping track of signs is a tricky business in supersymmetry.

$\int (a + b\theta)d\theta = b$.

Integration results in picking the last term in the expression.

How do we start describing a process in supersymmetry?

It always begins with choosing a field, building a Lagrangian, finding constraints, finding equations of motion, scattering process amplitudes and particle spectra.

You may have guessed that in supersymmetry, field is the super field and there is a super Lagrangian and finally a super particle.

We start with a super field. It depends on the usual space and time coordinates x^μ but also on θ coordinate. θ can be thought of as the super space. It is difficult to visualize but the properties help us to get the desired mathematical results

Super field $\Phi = \phi + \psi\theta + \bar{\Psi}\bar{\theta} + D\bar{\theta}^2\theta^2$ .... Here θ has 2 spin indices so θ^2 is allowed.

The super field has boson ϕ and fermion Ψ fields along with their conjugates represented by bar over them mixed with θs. It is basically a guess work and after manipulation we get the desired equations.

The next step is to invent generators so that we can see how this super field changes with change of coordinates. After all our ultimate goal is to understand how things change. The generators of coordinate transformation are called super generators.

$\theta = \theta + \varepsilon$ (small change)

$\bar\theta = \bar\theta + \bar\xi$ (small change)

The change in space time coordinates is a bit complicated.

$X^\mu = x^\mu - i\xi\,\sigma\,\bar\theta - i\bar\xi\,\sigma\,\theta$

This is how coordinates change in supersymmetry. The generators of these coordinate changes are called Q generators. I will omit indices in the equations for simplicity. σ is Pauli spin operator.

$Q = \dfrac{\partial}{\partial\theta} + i\sigma\bar\theta\,\dfrac{\partial}{\partial x^\mu}$

$\bar Q = \dfrac{\partial}{\partial\bar\theta} + i\sigma\bar\theta\,\dfrac{\partial}{\partial x^\mu}$

So, a small change in Φ is accomplished by Q's acting on it.

$Q\Phi$ or $\bar{Q}\Phi$

Here's a way to think about these generators. What are the generators to get a room full of people empty?

The answer in this case would be Q = fire alarm or bad smell!

We need some constraints. The constrained super field will be easier to handle.

Here's a constraint $\bar{D}\Phi = 0$

It means when this operator acts on Φ, nothing happens. These constrain operators are called super covariant derivatives.

$$\bar{D} = \frac{\partial}{\partial \bar{\theta}} - i\sigma\theta \frac{\partial}{\partial x^\mu}$$

$$D = \frac{\partial}{\partial \theta} - i\sigma\bar{\theta} \frac{\partial}{\partial x^\mu}$$

Note this looks very similar to Qs except with a sign change. Ds anti commute with Qs. $[Q\bar{Q}] = -[D\bar{D}]$

The other way to think about this is if Q acts on Φ then $\bar{D}$ acts, nothing happens.

$\bar{D}(Q\Phi) = 0$

So, we have to modify Φ in such a way to satisfy this constraint.

This new Φ will no longer depend no simple x^μ, θ and $\bar{\theta}$.

$\Phi(y, \theta)$ where y is $x^\mu + i\theta\sigma\bar{\theta}$. This is called a chiral super field.

It can further be expanded like

$\Phi = \phi(y) + \psi(y)\theta + F\theta^2$

Similarly, we can make complex conjugate of Φ by change of sign of y. Φ^2, Φ^3 like combinations can also be formed.

Let's begin by writing the super Lagrangian, sometimes written as Λ

Super Lagrangian Λ = Kinetic – Potential

Kinetic term = $\int \overline{\Phi}\Phi d^4x\theta$

The integration over all space time and θ takes place.

Now we want y dependence to change back to x, that can be done by shifting by $-i\overline{\theta}\sigma\theta$, this will result in y ($x^\mu + i\overline{\theta}\sigma\theta - i\overline{\theta}\sigma\theta$) of Φ to become $\Phi(x^\mu)$. It will also shift the other $\overline{\Phi}$ in the kinetic term by $2i\overline{\theta}\sigma\theta$. Then we eliminate $\overline{\Phi}$ dependence of this new variable back to x^μ. All these manipulations are to make everything depend on x^μ only.

The last step is to do integration and we use the tricks of Grassmann variables like picking the last term of the integral.

The details of math are onerous, but the upshot is that by doing all these manipulations, ywe get an equation which looks like

$\partial\overline{\phi}\partial\phi + \partial\overline{\psi}\sigma\psi + F^*F = 0$

Did you see the magic happen?

From an abstract superfield, super Lagrangian and super constraints, we derived an ordinary Lagrangian which can yield Klein Gordon(boson) and Dirac(fermion) equations. F is called an auxiliary field and it can be absorbed later into other variables.

The next step is to add super potential. This is where interactions and mass terms are found and there is lot of flexibility in what can be added. Things like $m\Phi^2 + m\Phi^3$ are good candidates. Recall each $\Phi = \phi(y) + \psi(y)\theta + F\theta^2$ and when we multiply this onto itself, we get different terms. The terms like $mF\bar{\phi}$ is a boson mass term, $m\bar{\psi}\psi$ is mass term for fermions. We can multiple potential with coupling factor like g, then we will have terms like $g\phi\bar{\psi}\psi$, with bosons and fermion interacting e.g. one boson scattering to two fermions. The other terms like $g\phi^2$ will represent boson scattering. The point to note is bosons and fermions have the same mass and coupling constant. This is what we wanted so that different boson and fermion loops can cancel to avoid infinities.

To solve the auxiliary field F and eliminate it from equations, let's set super potential $W = \Phi^N$. If we multiple the super field N times and collect F containing terms, we will find there are $F N \Phi^{N-1}$ terms or $F \frac{\partial W}{\partial \phi}$ terms. Similarly, there are complex conjugate $F^* \frac{\partial W}{\partial \phi^*}$ terms.

Let's collect all F terms and solve it.

$$F \frac{\partial W}{\partial \phi} + F^* \frac{\partial W^*}{\partial \phi^*} + F^*F = 0$$

The solution is $F = \frac{\partial W^*}{\partial \phi^*}$ and $F^* = -\frac{\partial W}{\partial \phi}$

And putting back these solutions back into the equation, we get an ordinary potential, not super potential $V = F^*F = \left|\frac{\partial W}{\partial \phi}\right|^2 = 0$

This means vacuum has no energy. That's well and good but it also means bosons and fermion super partners would have same masses. This is not experimentally true as no super partner has ever been observed. This means that super partners are heavier than usual particles but not too heavy. So, there is a slight splitting of mass levels. But what would cause that?

Now we are entering an even more mathematically difficult subject, Super symmetry breaking.

If $\left|\frac{\partial W}{\partial \phi}\right|^2 \neq 0$ or F is non-zero, that would give vacuum a positive energy. This would also mean that F field like a magnetic field would cause different effects on bosons and fermion super partners and split mass levels.

A positive vacuum energy produces gravity and our poor universe is like a ball on the top of a hill.

Our universe will roll down the hill and spread out in all directions, explaining accelerating and expanding universe. This is the origin of Dark Energy which is pulling universe apart. However, the vacuum energy has to be very small to expand the observed universe.

There are many models to make F non-zero.

$W = X\left(\frac{\lambda}{2}\Phi^2 - \mu^2\right) + m\Phi Y$

This has non-zero solution for $\left|\frac{\partial W}{\partial \phi}\right|^2$. This is known as O'Raifeartaigh model.

The super symmetry has to produce the Standard Model, the most successful physical theory to have any credibility. The model that mimics it is called Minimal Supersymmetric Model (MSSM).

It involves more complicated Kinetic and Potential terms. There is also a term to break the symmetry as well.

L = ∫ L (kinetic/gauge) + ∫ L (Yukawa/interaction) + ∫ L scalar (F, D) +∫ L (symmetry break)

There is another potential besides W, called V, called super vector potential. It is real unlike W and its auxiliary field is called D. This field goes into the gauge equation. There is doubling of all the particles of standard model and there are 2 Higgs fields. Bottom line is it is mind bogglingly complicated!

The symmetry breaking is done softy which is done in the so-called hidden sector.

The hidden sector means hidden super particles do not interact with Standard Model force mediators like gluons, photons and remain hidden from the usual Standard Model. The soft symmetry breaking means that splitting of symmetry should not be too strong that it causes divergences or infinities as this was the motivation for the super symmetry in the first place. More specifically soft symmetry breaking happens at low energies. The high energy scattering is not affected, and thus ultraviolet divergences are avoided.

The hidden sector mediates this symmetry to the visible sector (Standard Model particles) breaking through gravity and gauge interactions. The gravity involved is called super gravity or SUGRA. I have neglected to mention that supersymmetry is fondly called SUSY. The boson super partners have s as a prefix like selectron for super partner of electron, sup quark for up quark. The super partners of fermions have-ino at the end e.g. higgsino, gluino etc. Who says physicists can't have fun with names?

This model also predicts decay of proton. To prevent this, we invent R parity in the model, so that standard particles have even parity and super partners have odd parity. This is like a charge that is preserved in interactions. This way unwanted decays are prevented. This parity also leads to the lightest super partner to be stable and a candidate for Dark Matter.

I am sure you must have heard about Dark Matter. It was postulated when cosmologists calculated the velocities of outer rim around galaxies and found that their value cannot be explained by the visible mass of the galaxy. So, there must be hidden matter making galaxies heavy to give outer galaxies their velocity. The stuff that makes dark matter is a matter of speculation as its largely hidden from experiments and interactions with visible particles.

The super symmetry improves on the Standard Model by explaining Higgs mass, but it also leads to problems. The parameters that are put in by hand in the Standard Model are close to 20 but in super symmetry, they reach 100. This feels very unnatural as if too much tuning is going on to get the right results. The fact remains that so far absolutely no evidence exists for the super partners. The mass of Higgs as experimentally detected, is not supportive of the existence of super partners. The hope that super partners will be seen at Large Hadron Collider is receding as years go by and no evidence emerges on their existence. I am reminded of a story told by Dr. Abdus Salam in a video made in 1986. He was giving lecture on supersymmetry in 1974 at Miami. Paul Dirac was in the audience and as usual he kept quiet. When Dr. Salam asked him what do you think about this mathematically beautiful supersymmetry theory? Dirac softly told him that if supersymmetry was true, it would have been discovered long ago. It appears that Dirac 's instinct may be true as decades have gone past and no experimental evidence exists to support it.

Let me give you a layman's perspective on this. We as humans belong to different races, ethnicities, religions etc. but fundamentally we are all the same. How do we know that? DNA of course, we all share the basic human genome. The discovery of DNA no doubt was a crowning achievement. It explained the symmetry behind all the diversity in human beings. The more difficult thing is to go from DNA and explain each and every diversity in humans. Why someone is taller, thinner, has certain hair or skin color etc.

This is the story of super symmetry. Once we show that bosons and fermions are alike, in the sense that they have same mass and coupling constants. Then to go backwards and show the experimental differences that are seen is even more complicated. This is what SUSY breaking does. The non-zero vacuum energy to explain SUSY breaking can be thought like this.

Let's say standard people live on the mainland. Then there are people who live on a remote island. The things on the island remain hidden as its far away with little interaction with the mainland. If there is abnormal radiation near the island, it can cause mutations and different diseases in the island people. The sick people break the symmetry with the healthy people. This is like saying that the non-zero vacuum energy manifesting via F field interacts with the hidden sector of super particles and breaks the symmetry causing mass splitting. The diseases can be transported to the mainland through ships and air travel. This is like our hidden sector transmitting symmetry breaking to the visible standard model particles.

This is of course an absurd analogy but remember the key point that the hidden sector symmetry breaks via non-zero vacuum energy and it is passed on to the Standard Model particles through gravity.

I will end this chapter by giving another trivial analogy. The concept that the relationships are made in heaven suits supersymmetry very well. No one is lonely. God has made partner for everyone. It brings balance to life. If one partner is extrovert, other is introvert. The introvert partner cannot share the same space with another introvert. $\{I, I\} = 0$, this is our anti commutation relation. The extrovert partner is extravagant, and the introvert is thrifty.

Action $(X, \theta) = \int$ extrovert partner $+ \int$ introvert partner

X= where they go, who they meet

θ = personality space, how personalities influence interactions

Action is the limited budget

The extrovert contribution (boson loop) = buy this + that + this.........

We need to renormalize it so that it stays in the budget like renormalizing Higgs mass.

This is the role of the introvert.

The introvert contribution (fermion loop) = - No, you can't – No, not this….

Limited budget= buy this + that + this……….- No, you can't – No, not this….

So, each partner compliments each other and they avoid financial ruin.

I will leave up to you to decide if my analogies are more absurd or renormalization mathematics is more palatable to you.

String Theory

If you are reading this book, I am sure you must have heard of String Theory. It is all the buzz in the media and popular books, that describe it as the Ultimate Theory of Everything. Getting away from cliché, let's describe the theory as it is. The most unique aspect of this theory is that particles are not point but string like objects. This basic fact has let to tremendous mathematical exploration of the subject. This theory combines Einstein's General theory of Relativity and Quantum Mechanics, which has been the failure of standard field theories. The fact that gravity naturally comes out of this theory along with the other particles is very tempting. On the flip side, string theory requires extra dimensions and there are many versions to choose from, which is annoying. The fact remains that experimental results confirming string theory remain elusive, which has resulted in this subject becoming controversial. There are physicists and then there are string theorists!

Historically, this subject came to life in late 60's while trying to explain the strong nuclear force that binds protons and neutrons. This was before Quantum Chromodynamics (QCD) was discovered and once QCD was successful in explaining the strong force, string theory remained in oblivion for many years. The early hints of string theory came from experimental results that m^2 (mass) is related to angular momentum and if we plot a chart of m^2 vs angular momentum of various hadrons, we get a straight line. These particles on scattering made special trajectories called Regge trajectories. The Italian Physicist Veneziano came up with a formula for the scattering amplitude of the mesons, explaining some of these experimental properties. Let's look at that formula

$A = \frac{\Gamma\Gamma}{\Gamma}$ (s,t) where Γ (gamma function) is n-1!(factorial)

Gamma function is a kind of factorial. The usual factorials only work for positive integers but with gamma function, complex functions can be included. The factorials are used to arrange or combine things. If you want to arrange ABC letters, the answer is 3! which is 3×2×1= 6. Likewise, you can think of gamma function in this context as arranging incoming and outgoing particles with various momenta. Here S channel refers to 2 particles annihilating and forming other particles through intermediate virtual particles. T channel refers to particles exchanging virtual particles and then undergoing change in their momenta. Let's look at the pictures as picture is worth a thousand words!

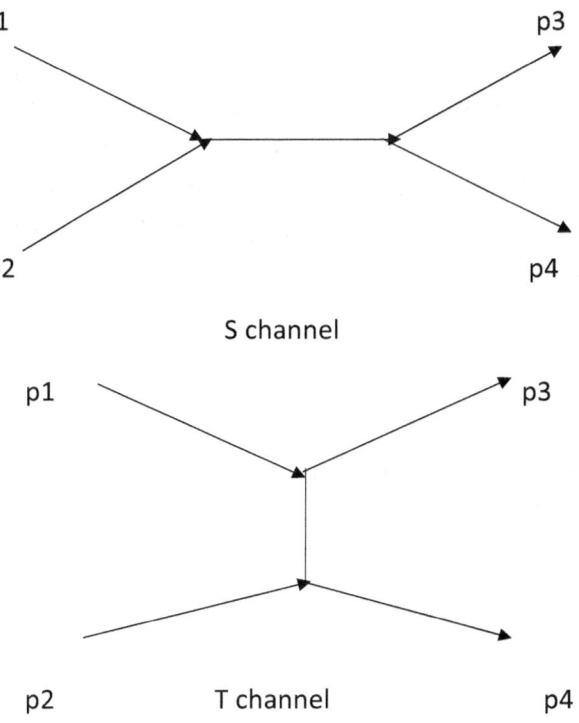

In QFT, s and t channel processes have separate amplitudes and we have to add them to get the final probability of scattering. But in Veneziano amplitude, the neat thing is that s and t channel processes were coded into a single formula and so we don't need to separate them. This led to the conjecture that may be particles are like strings and not point like. Something like -

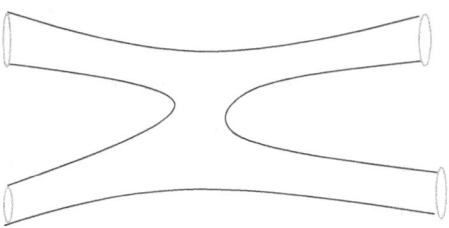

This is a world sheet in the language of string theory and it replaces the line like Feynman diagrams from QFT. The incoming strings come and join then go out as outgoing strings, describing a scattering process. The early versions of string theory predicted weird particles and 26 dimensions! No wonder, physicist community was very skeptical, and it took decades for string theory to gain prominence.

Formulation

The strings can be open or closed. The area they cover form the world sheet. We need some parameters to describe a string.

The shape of the string is arbitrary. Various shapes or vibration modes represent different particles.

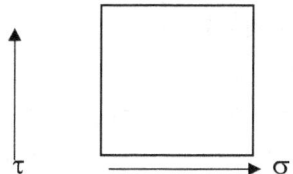

τ is time like dimension and σ is space like dimension. We are free to choose them as these are arbitrary dimensions. These are not real dimensions like x, y, z. These are mathematical parameters to describe a string. Sometimes they are written as σ^α where α runs from 0 to 1. τ =0 and σ = 1 in this notation.

Finding the area of world sheet is a good approximation for the interaction of strings. Finding the Lagrangian and minimizing it to get the action(S) is the standard physics recipe to describe any system.

$$S = -T \int d\tau d\sigma \sqrt{-\det \gamma} \quad \text{where det of } \gamma \text{ is } \sqrt{-\left(\frac{\partial x}{\partial \tau}\right)^2 \left(\frac{\partial x}{\partial \sigma}\right)^2 + \left(\frac{\partial x}{\partial \tau} \cdot \frac{\partial x}{\partial \sigma}\right)^2}$$

x has x^μ index and it can run based on number of space and time dimensions. T is tension of the string. γ represents how these world sheet parameters interact with curved space background they are embedded in. Starting with a tiny segment of the area and then doing integration gives the whole world sheet area. This is known as Nambu- Goto action.

There is another way to formulate the action known as Polyakov action. This action removes the nasty square root.

$$S(\text{Polyakov}) = \frac{-T}{2} \int \partial\tau \partial\sigma \sqrt{-\det h} \, \partial x^\mu \partial y^\nu g$$

Where h is metric of the world sheet and g is the metric of the space it is embedded in. It is like saying a cup has a metric and the room it is embedded in has its own metric, which represents the curvature of space time. Don't take this analogy too seriously as we cannot separate cup from the room or the space we live in as we use the same space-time dimensions to measure the cup as well as the room.

But this is math, so you can do anything! The metric of the world sheet is a mathematical concept only. It can be made flat and removed from the equations by suitable tricks as the metric is completely arbitrary.

Next, we find equations of motion, various symmetries, constraints and finally we do quantization where we get various particles predicted by the theory. This involves complex mathematics to describe each step. We will only do some of them briefly to get an idea.

The equations of motion can be quite nasty. So, we look for ways to simplify them. Light cone gauge is an important trick to simplify equations. We boost the particle along one direction of motion so that momentum in that direction becomes much bigger than the other directions. This way remaining directions can be treated non relativistically, which means simpler equations, by ignoring special theory of relativity.

$$E = \sqrt{p^2 + m^2}$$

taking speed of light c = 1.

If p is very big as compared to m then $p\sqrt{1 + \frac{m^2}{p^2}}$

The square root term can undergo binomial expansion which says $\sqrt{1+\Delta} = 1 + \frac{\Delta}{2} - \frac{1}{8}\Delta^2 + \frac{1}{16}\Delta^3$ .... where Δ is a small number as p is much bigger than m.

We will only focus on first and second order terms only.

The expansion will give $p + \frac{1}{2}\frac{m^2}{p}$.

Since momentum is huge in the chosen direction, it can be regarded as constant.

The take away point is, $E \sim m^2$

Why it's called light cone gauge?

Well in modern versions of the theory, one space and time direction are combined to build a new coordinate.

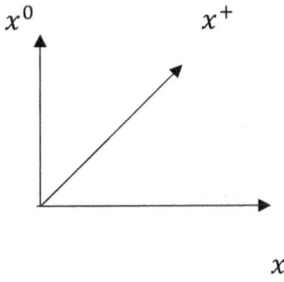

$x^+ = \frac{1}{\sqrt{2}}(x^0 + x^1)$

The new coordinate is at 45 degree or on the boundary of the light cone.

Similarly, we can form $x^- = \frac{1}{\sqrt{2}}(x^0 + x^1)$

$2dx^+ dx^- = (x^0)^2 - (x^1)^2$

Thus, the space time distance $ds^2 = (x^0)^2 - (x^1)^2 - (x^2)^2 - (x^3)^2$

is replaced by light cone coordinates to do further simplification of equations. We will leave it at that.

We will only be doing the rudimentary string theory, which is called the bosonic string theory. It only produces bosons or forces. No fermions are produced. It has several problems but for our purpose it provides a good overview of string theory. The modern versions are too complicated to include here.

Let's quantize, shall we. Since we are dealing with non-relativistic QM, our friends – annihilation and creation operators of harmonic oscillator will do the trick. After all vibrating strings are like harmonic oscillators. The strings vibrate in different modes, meaning how fast or slow are they vibrating. Each mode represents different energy and particle. The ladder operators help us navigate from one mode to another. This is called quantization.

$$L^+|\psi_{ground}\rangle = m_g^2 + 1$$

L^+ is the creation operator

$$L^-|\psi_{ground}\rangle = 0$$

L^- is the annihilation operator. Obviously, it cannot lower the energy of the ground state.

We can use separate operators for x and y directions and get new state with 1-unit higher energy than the ground state level. The important thing to know is that this quantization in x and y direction is like polarization of photon. Photon moves in say z direction and electric, magnetic fields oscillate perpendicularly in the other directions. Since photon is massless, no polarization in the direction of motion is allowed. Similarly, we see that if particle is moving in z direction with huge velocity then its momentum can be considered constant. We are creating higher energy states in x and y directions with the help of ladder operators. The interpretation is that we are creating polarization states in two directions perpendicular to the direction of the motion of the particle. This means the ladder operator acting on the ground state creates a massless particle with spin 0. This means we have created a photon. This is good.

If $m_g^2 + 1 = 0$ then m_g is negative, that's not good. There is a name for this negative mass particle, Tachyon. This particle has never been observed and it did raise many questions about the relevance of string theory in early years. Modern string theory can remove these particles by suitable constraints.

Let's look at some of the constraints used to get the good states. In case of closed strings, we assume that right moving and left moving modes on the string should have the same energy as the direction is completely arbitrary. It's like our clock, the fact it runs "clockwise" is a convention. So instead of using individual operators, x and y direction operators can be combined e.g.

L can be Right or Left moving (+1 or -1)

M can be Right or Left moving (+1 or -1)

$L_1^+ M_{-1}^+ |\psi_{ground}\rangle$ is allowed and has 2 units of energy

We can do linear combinations of L and M as well. It is possible to get a massless particle with spin 2. The massless particle with spin 2 is called the graviton. This is a big deal as gravity now emerges naturally in this theory. There are other particles that come up in the spectrum. None have been seen in experiments.

Extra Dimensions

Science fiction is pale in comparison to what physicists propose in their theories. The early version of the string theory required 26 dimensions. Now that is some wacky stuff! Let's explore the reason for these extra dimensions.

The string can be divided into n parts. Each part vibrating with various modes. The division is arbitrary, n can be a huge number This leads to a problem as E~ m^2 and we found earlier that m^2 = -1, so energy of the ground state should equal to -1.

$$E \sim \sum_{n=1}^{\infty} = -1$$

This is like saying 1+2+3+4......= -1. As we sum up this infinite series, we get larger and larger sum. So, how do we make it equal to -1?

Let's take some math diversion to learn about the fascinating problem of summing infinite divergent series. This problem in mathematics has been solved in various contexts in the past and physicists just borrow the results to fit in the theory.

If I tell you that the sum of the infinite series 1+2+3+4......= - 1/12, would you believe it. This is what a young man living in a remote village of India wrote in 1913 to a world-renowned mathematician at Cambridge. The man was Srinivasan Ramanujan, who later became a legend in the field of mathematics. He made ground breaking discoveries on his own without any formal education and was recognized through his letters by Cambridge professor GH Hardy. There are plenty of YouTube videos about this and his fascinating life was made into a movie "The Man Who Knew Infinity".

We need to know about complex analysis and analytic continuation of functions to understand this problem. Don't worry, I will simplify this scary looking math jargon.

Riemannian Zeta function is another way to evaluate summing of infinite series, where we can sum up complex numbers as well.

$$\varsigma(s) = \frac{1}{1^s} + \frac{1}{2^s} + \frac{1}{3^s} + \frac{1}{4^s} \ldots\ldots$$

This sum is defined only for s> 1. When s is -1 then we get 1+2+3+4....... and then the answer is – 1/12. Since s is only defined for s>1 then how did we get the answer for -1. The answer is continuation of analytic function. We just extend the domain of the function to other regions. Of course, there are certain rules and you should realize that -1/12 is an approximation only. The rules involve extending the function to the new domain where it can still be analyzed i.e. differentiation and integration is well defined. The curves of a function with different values of x, make angles with one another. In the new extended domain, different curves should also make similar angles. This is known as conformal symmetry.

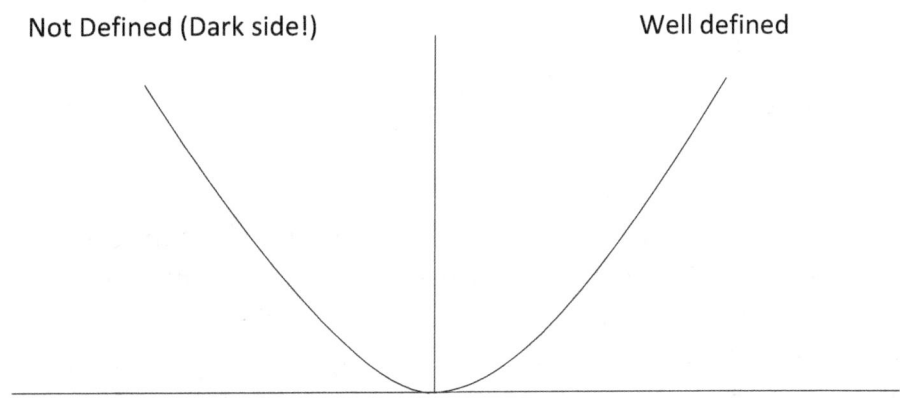

Analytic continuation of function

When in Rome, do as Romans do is used in reverse for the analytic continuation of functions. It's like when going to an alien land, we have no idea about how to interact with aliens. We use our common sense from our own familiar and defined culture and hope it all works out! This is what the crew of starship enterprise does, continuously extending their biases to alien lands!

Ok, back to the string theory. So, we know that the sum of our oscillators is -1/12. This is only one direction, adding the other direction takes it to -1/24. The remaining z direction that we boosted the string along to do the light cone gauge and the time dimension makes a total of 26 dimensions.

How to make sense of these 26 dimensions?

I will give another trivial analogy to help you ponder over it. Let's say you weigh 60kg and need certain calories per day to maintain this weight. You can eat whatever is available in the grocery store, coffee shops, restaurants etc. How do we mathematically formulate it?

Weight= $\sum$ bread+ cheese+ juice +meat+ fruits……= 60kg

In real life, you can be observed to know exactly what you eat so that each calorie is counted. But what if like strings and elementary particles, we have no idea what you eat and how much. In that case this sum will be infinite as all possibilities are equal and have to be added to the sum. The only way to make sense of this infinite sum is to put some constraints.

1. Cut off – the number of calories you can eat is limited by money and how much time is available in a day to eat. This is like our renormalization procedure in QFT.

2. The other possibility is that indeed you eat a lot, but that weight goes in extra dimensions that cannot be measured by the weighing machine. So, 26 extra dimensions are a perfect weight maintenance program. Whatever you eat, weight remains 60kg.

What to do with these damn extra dimensions?

We shrink them. They are made so small that no one can see them, but they are available mathematically to fine tune our theory. How convenient!

The process of shrinking extra dimensions is called Compactification. It is a mathematically dense subject, but we will look at it geometrically, which is a bit easy to understand. It is like mapping. We all see how our 3-dimensional world is mapped onto a 2-dimensional atlas. Likewise, extra dimensions can be mapped onto a reduced dimensional surface and magically the extra dimensions disappear.

All the points on 3d surface are identified on the 2d surface. The mapped surface is made flat mathematically, known as Ricci flat as many equations are simplified.

More Constraints

The open strings can be constrained based on how their ends behave. The strings can attach at certain points, it is known as Dirichlet boundary condition. The strings can be relaxed at their ends meaning no tension at the end of the string or in other words the derivative vanishes. This is the Neumann boundary condition.

Dirichlet condition $\delta x = 0$

Neumann condition $\dfrac{\partial x}{\partial \sigma} = 0$

These conditions apply at the string ends which run from 0 to π for open strings.

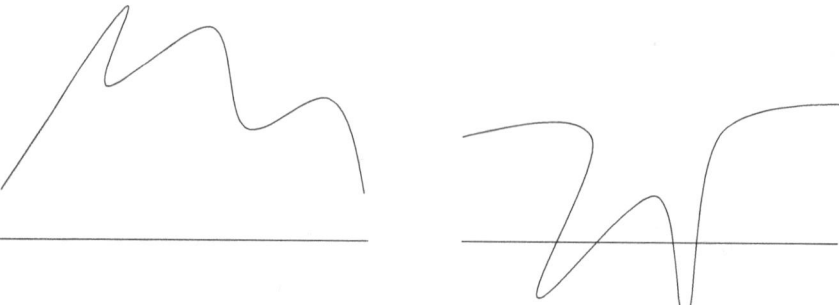

Dirichlet (strings are fixed)

Neumann (strings are free at the ends)

On a compacted surface, strings can be considered either winding onto or traveling along the compacted direction.

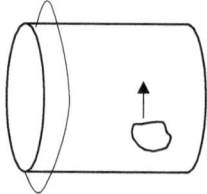

Cylinder represents compacted surface. The string can travel around it or be wound around the surface.

It turns out that if the surface is of radius r, then Energy is $\sim r$ for the winding case and 1/r for the traveling case. If r is very big, then winding case will have large gap in energy spectra but traveling string will have very closely placed energy levels. The situation reverses for 1/r case. The upshot is that winding or traveling strings have a duality, meaning they both represent the same thing.

This is known as t-duality. This duality led to the discovery that many string theories with different formulations are exactly the same as some represented winding, others represented traveling along the compacted directions. This was a relief as total number of string theories were getting out of hand. You could find a string theory beneath every rock!

Branes

The Dirichlet condition of strings fixed to a surface begs the question, what is this surface? This hyper surface is called a D-brane. It can have various dimensions. D-0 is a point, D-1 is like a string, D-2 like a membrane etc.

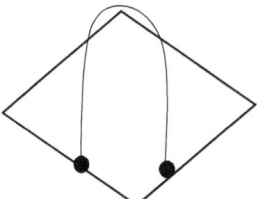

Different branes can interact with each other and exchange strings. This can be used as a model for explaining QCD where each brane can represent red, green and blue color and strings connecting each brane can be thought of as gluons.

The brane can be of any size, even bigger than a galaxy or universe for that matter.

M Theory

The extra dimensions and branes give theory tremendous flexibility as well as difficulty. The extra dimensions can help explain why gravity is so weak as compared to nuclear forces. Maybe gravity leaks out to extra dimensions. The problem with flexibility is that it led to many string theories to be proposed, a problem of plenty. The term M theory was coined by Edward Witten. It refers to a yet fully undiscovered theory that lies beneath various string theories. We have seen how t duality connects some string theories, but finding other connections remains a work in progress. The M theory has reduced the number of extra dimensions from 26 to 11 dimensions, what a relief! The string theory discussed thus far contains no fermions. The early version of the string theory as discussed before is termed as bosonic string theory. To add fermions, we have to add them as bosonic super partners and then it becomes a superstring theory. You can think of various string theories as different states of America. You know there is something that unites them, but they look different as well. The M theory is like whole of United States of America. The physicists are like early explorers, still far away from fully exploring it.

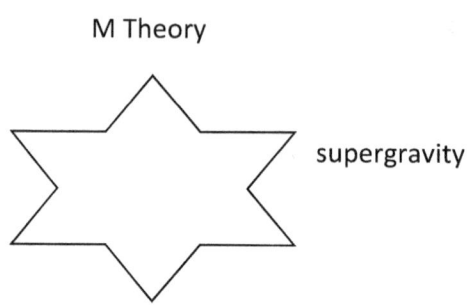

AdS/CFT duality

Before we wrap up the chapter on string theory, I want to briefly touch on AdS/CFT duality as it is the topic of special interest in the field of string theory and we should know something about this. This conjecture as proposed by Juan Malcadena in 1997 has created quite a buzz in the physics community. The detailed mathematics is onerous, but I will try to explain it qualitatively. AdS stands for anti-de Sitter's space which is a negatively curved space. Einstein's General Theory of Relativity tells us that space time is a dynamic place and it can have a positive or negative curvature. The string theory in this higher dimensional negatively curved anti de Sitter's space is equivalent to quantum field like theory in the lower energy limit. The field theory lives on the boundary of this space time and it encodes all the information of the inner space time like hologram. The field theory that works with this kind of arrangement is called N =4 Super Yang Mills theory. If super symmetry algebra was not difficult enough, where N=1 refers to one super symmetry, meaning every boson has a fermion partner. In N =4, there are 4 super symmetries. The advantage of more supersymmetries is that this theory is conformaly invariant. This is like saying that if Newton's laws of motion are applicable in the real 3D world, then if these laws remain unchanged as we apply them in a 2D map of the world, the laws are conformaly invariant. This has many applications as one can use QFT techniques to study more complicated string theory and get the same information. Black hole cosmology uses this conjecture to study various properties of these fascinating creatures of this Universe.

This was a very short review of string theory. It is a mathematically challenging subject, but I hope you got some idea about the working of the theory. We did not go into details of gravity. This was deliberate as without knowing Einstein's general theory of relativity, it is not possible to have a discussion about gravity. Let's leave that for some other time.

String boxing

Physicists are in demand everywhere, from Wall Street to Hollywood. Let's see how a string theorist analyzes a boxing match.

1. He will look at the parameters of the boxers and the match. This would mean height, weight, venue of the fight etc. This is like σ and τ of the string.
2. He will look at the strength and weakness of the boxers. What they have got? How much energy and stamina are there to throw punches and play all 12 rounds. This is like writing a Lagrangian. This Lagrangian may contain metric of the boxer and the ring meaning how flexible the boxer is in the match.
3. He will get familiarized with the constraints. This would mean rules like staying in the ring, cannot hit below the belt etc. These are like constraints in the string theory e.g. Dirichlet, Neumann boundary conditions.
4. He will develop a strategy for boxing like how to throw punches and make moves in the ring. It will depend on the strength of each boxer. This is like equations of motion derived from the Lagrangian.

5. He will further simplify his strategy by drawing it on a piece of paper. This is like compactification or shrinking of dimensions and using flat space to do planning.
6. He will analyze the invariances and symmetries of the match. It means things like which color clothes are worn, who enters the ring first etc. They have no bearing on the match. This is like gauge invariance in field theory that help us to construct various scenarios for describing the match, but do not affect the outcome.
7. He will break down the fight punch by punch. This is the quantization procedure.
8. In the end, he will have the information to study and predict which punches will result in a knockout. He would know the entire scattering process and spectrum to make experimental predictions.

Our physicist is very happy that he can know predict the next heavy weight boxing champion, who will be the next Ali or Tyson. He expects to be rewarded. He calls up the boxing federation for a reward. The federation member says I know your predictions are stellar, but they work only on paper and in video games. Unless you show results in real life, we are going to hold the reward as Nobel committee has done so far for string theorists.

Conclusion

We have come far. We started with elementary quantum mechanics and ended with the M-theory. There is a lot to digest. I do not expect you to know every equation or each step in the derivations. I would advise you to focus on the concepts. There is no substitute to intuition and visualization. Try to think over how these complex concepts apply to daily life. It means coming up with analogies, no matter how absurd they are. The goal is to enrich our life with the tremendous wealth of knowledge that physics has to offer, without worrying about exams and assignments. Once you feel confident and familiar with the working of quantum mechanics, you could attempt to read standard texts and take the quantum leap to become a physicist.

References

I am extremely grateful to the sources that provided invaluable information in learning physics. I have listed the important sources for your review and tried not to miss anything inadvertently. If there is any mistake in the list, please let me know.

1. Griffiths, D (2017) Introduction to Quantum Mechanics, 2nd edition, Cambridge University Press, New York. It is THE book to study quantum mechanics as an undergraduate. Everything in the book is exemplary and first class. It is the standard textbook across many universities and its notation is widely adopted.

2. Shankar R, (1994) Principles of Quantum Mechanics, 2nd edition, Springer Science+ Business Media, New York. An advanced text that covers more topics than Griffith's. It needs a strong mathematical background and notation is higher level.

3. Griffiths, D (2010) Introduction to Elementary Particles, 2nd revised edition, Wiley-VCH, Weinheim. This is the best introductory text book for particle physics in my opinion. It is a difficult subject, but Professor Griffith makes it reasonably accessible. It's no surprise that his notation is widely adopted.

4. Susskind L, Friedman A (2014) Quantum Mechanics, The Theoretical Minimum, Basic Books, Philadelphia. If you are familiar with Professor Susskind through the You Tube series, this book is a good supplement. It can be used as a stand lone book as well where it goes beyond popular books to explain the mathematics behind quantum mechanics in an engaging way.

5. Susskind L, Hrabovsky G (2013) The Theoretical Minimum, Basic Books, Philadelphia. A great introductory text for classical physics and concepts like principle of least action, which are precursors to learning QM.

6. Harris, R (2008) Modern Physics, 2nd edition, Pearson Addison-Wesley, San Francisco. An excellent introductory text. It is focused on practical aspects than a dedicated QM text, but clarity of writing is commendable.

7. Lambourne, R (2010) Relativity, Gravitation and Cosmology, Cambridge University Press, New York. An excellent introduction to special theory of relativity and relativistic notation.

8. Serway R, Beichner R, Jewett, J (2000) Physics for Scientists and Engieeners, 5th edition, Harcourt College Publishers, Orlando. A classic physics textbook. Refresh your classical physics and there is an excellent chapter on special theory of relativity.

Online Resources

Information is increasingly available online. It is not possible to only use books for learning physics. Online tools are critical in self-study. I don't think it's a surprise for you that every topic gets googled and first checked on Wikipedia. I will list quality sources that I found useful. Since website addresses get changed, I have included the description of the source so that it can be easily searched.

1. Perimeter Institute of Theoretical Physics at Waterloo, Ontario has an excellent outreach program. You can find lots of videos on quantum mechanics and advanced topics at perimeterinstitute.ca.

2. University of California at San Diego has several online topics on quantum mechanics. Search for ucsd.edu quantum mechanics.

3. University of California at Berkeley- search for Professor Umesh Vazirani. He has lots of excellent lecture notes on quantum computing on his personal Berkeley page.

4. Stanford University has YouTube channel. Professor Susskind has CME video series especially for public, covering topics like quantum mechanics, particle physics, special and general theory of relativity and string theory. These lectures are a treasure. Professor Susskind has a dynamic personality and he is very engaging while teaching physics.

5. MIT has an open course ware website at ocw.mit.edu where there are extensive lectures and videos on topics like quantum mechanics, particle physics and string theory.

6. University of Cambridge high energy physics group website at hep.phy.cam.ac.uk has excellent articles on particle physics under Professor Mark Thomson lecture courses.

7. Arxiv.org is a repository of publications, run by Cornell University. It has some excellent articles. Introduction to the MSSM by Sudhir Vempati and TASI 2008 Lectures: Introduction to Supersymmetry and Supersymmetry Breaking by Yuri Shirman are excellent articles to get the basics of Supersymmetry.

8. Khanacademy.org is an excellent online source to learn mathematical topics like calculus, complex functions and series expansions etc.

Index

A

Adiabatic Approximation 303

Aharanov-Bohm Effect 310

Alpha Particle 280

Angular Momentum 139

Azimuthal Equation 145

Airy Solutions 286

B

Band Theory 215

Bell's Inequality 181

Blackbody Radiation 234

Bloch Theorem 210

Bohr's Radius 136

Born Interpretation 41

Boson 204

Bosonic Statistics 223

Bra-ket Notation 43

C

Calculus 11

Cauchy Formula 330

Clebsch-Gordon Coefficients

C

Commutation 73

Complex Numbers 20

Conservation Laws 27

CPT 427

Cross Product 140

D

Differentiation 12,14

Differential Equation 17

Dirac Delta function 46

Dirac Equation 373

Double Slit Experiment 32

E

Eigenfunction 60

Eigenvalue 60

Electric-Dipole Approximation 301

Electroweak Unification 433

F

Fermion 204

Fermi Surface 208

Fine structure 251

F

Finite Potential 96

Forbidden Region 98

Fourier Transform 67

Free Particle 64

Feynman Diagram 388

G

Group Velocity 70

H

Hamiltonian Mechanics 23

Harmonic Oscillator 104

Heisenberg Uncertainty 79

Helicity 382

Higgs Boson 343,444

Hydrogen 118

Hyperfine Structure 261

I

Identical Particles 202

Imaginary Numbers 18

Integration 15

Inner Product 44

K

Kronecker Delta

K

Klein-Gordon Equation 371

L

Lagrangian Mechanics 25

Lagrange Multiplier 226

Ladder Operators 115

Laguerre Equation 133

Legendre Function 124

Length Contraction 358

Leptons 339

Liouville Theorem 177

M

Majorana Fermion 381

Maxwell Equations 286

Mesons 410

M Theory 483

N

Neutrino 450

Noether's Theorem 27

Normalization 42

O

Operators 52,53,62,153

Outer Product 243

P

Path Integral 86

Particle in a Box 89

Partial Wave Analysis 317

Perturbation Theory 239

Phase Velocity 68

Photoelectric Effect 31

Poisson Bracket 85

Potential Step 100

Power Series 107

Proca Equation 383

Q

QCD 412

QED 394

QFT 438

Quantum Computing 186

Quantum Cryptography 195

Quantum Entanglement 179

Quantum Statistical Mechanics 216

Quark 340

R

Radial Equation 129

Recursion Formula 108

Renormalization 401

S

Schrodinger Equation 51

Singlet State 166

Spherical Coordinates 119

Spherical Harmonics 127

Spin 147

Standard Model 338

String Theory 470

Supersymmetry 457

T

Tensor Product 170

Triplet State 166

U

Uncertainty Principle 72

V

Variational Principle 265

W

Wave function 40

Weak Interaction 420

Weyl Fermion 381

WKB Approximation 274

Y

Yang Mills Theory 442

Z

Zeeman Effect 257